PLANT NAMES EXPLAINED

Botanical Terms and their Meaning

David and Charles

A DAVID & CHARLES BOOK
Copyright © David & Charles Limited 2005

David & Charles is an F+W Publications Inc. company
4700 East Galbraith Road
Cincinnati, OH 45236

First published in the UK in 2005
Reprinted 2006

A catalogue record for this book is available from the British Library.

ISBN-13: 978-0-7153-2188-1
ISBN-10: 0-7153-2188-9

Printed in Singapore by KHL Printing Co Pte Ltd
for David & Charles
Brunel House   Newton Abbot   Devon

Commissioning Editor Mic Cady
Art Editor Alison Myer
Production Director Roger Lane

Produced for David & Charles by
OutHouse Publishing
Winchester, Hampshire SO22 5DS

Compiler Julia Brittain
Editor Sue Gordon
Designer Dawn Terrey

Visit our website at www.davidandcharles.co.uk

David & Charles books are available from all good bookshops; alternatively you can contact
our Orderline on 0870 9908222 or write to us at FREEPOST EX2 110, D&C Direct,
Newton Abbot, TQ12 4ZZ (no stamp required UK only); US customers call 800-289-0963
and Canadian customers call 800-840-5220.

CONTENTS

FOREWORD

Any visit to a foreign country is enhanced by the possession of a little local knowledge. If you speak a few words of the language, it is always easier to get what you want. So it is with a visit to the world of plants: a basic understanding of botanical Latin makes the gardener, new or experienced, more comfortable in his understanding of the natives.

Latin names are nothing to be afraid of. They may be unfamiliar at first but, as soon as a few words are mastered, the desire for more knowledge will be kindled. The classical languages from which nomenclature is derived are the roots of many other languages. Once one moves on from the initial unfamiliarity, many terms will seem more logical and their meaning often becomes obvious.

Plant names tell many stories, both factual and imaginary. They are a fascinating catalogue created by those that discovered the plants and the botanists who tried to fit those plants into the infinite jigsaw of the plant world. The names given to the plants are valuable information to the

ARBUTUS UNEDO

VALERIANA PHU

gardener; often descriptive and cultural but also a historical record of where a plant fits into the history of our gardens.

Botanical Latin is a living, evolving language. Names change as we learn more about the plants we discover and cultivate. These name changes are often an annoyance to the gardener but a joy to the botanist who has come closer to fitting a piece of the jigsaw into the right position. The most magical property of the language is its international application, a passport for gardeners and plantsmen everywhere to communicate, regardless of their native tongues.

Plant names are fun. Modern cultivar names add a familiar dimension to the classical roots of the language. In our gardens they introduce a host of characters into the border, regardless of their place in history. You will find the famous, the infamous and the improbable sharing the same bed right outside your window.

Andrew McIndoe

MYRTUS COMMUNIS

INULA HELENIUM

ABOUT THIS BOOK

The framework of *Plant Names Explained* is an alphabetical listing, giving translations of many of the descriptive terms (specific epithets, see page 11) that form part of Latin plant names. Often an example of a relevant plant name accompanies the translation; many of these will be familiar garden plants. Some specific epithets (e.g. *angustifolius*, narrow-leaved) are more common than others, occurring in the names of a number of plants. These are picked out in the listing as '**Common descriptive terms**', each accompanied by several plant examples illustrating use of the term.

Once translated into English, specific epithets often reveal useful information about the plant such as its colour, native habitat or country of origin. For general interest and easy reference, the book features a series of '**What's in a Name?**' panels, each listing a selection of these Latin descriptive terms with their meanings. These panels are grouped by theme such as shape, plant uses, colours or habitat.

Genus names (see page 11) often derive from the names of notable people, or from Greek, Latin or other ancient languages. For example, *Aubrieta* is named for Claude Aubriet,

a French botanical artist, while *Chrysanthemum* is from the Greek words for 'gold flower'. The book features a series of '**Genus Names**' panels (pale green) explaining the derivation of some of the main genus names beginning with each letter of the alphabet.

Sometimes it is useful to look up the Latin name of a plant of which you know only the common name, for example foxglove, snowdrop or beech. Another letter-by-letter series of panels answers the question 'What's the Latin name?' Herbs, fruits and vegetables are listed in separate '**What's the Latin Name?**' boxes on pages 47, 169, and 113 respectively.

Modern plant names include a whole world of cultivar names (see pages 13–14). These can be useful, funny or just plain fascinating. Selected foreign cultivar names, such as *Philadelphus* 'Manteau d'Hermine' or *Prunus* 'Amanogawa', are translated in a series of alphabetical 'Cultivar names' panels: '**Foreign Expressions**'. A selection of the more quirky English cultivar names are highlighted in a similar alphabetical series called '**Strange but True**'. A third letter-by-letter series of panels is '**Personal Names**', featuring plants such as *Narcissus* 'Jenny',

Rhododendron 'David' and *Lonicera periclymenum* 'Sweet Sue'. Other cultivar names highlight people or places, artists or foods, or are named for special occasions such as birthdays or anniversaries (useful when thinking about gifts). Examples of these are listed in themed '**Cultivar Names**' panels scattered through the book.

For a full list of all the above alphabetic and thematic panels in the book, see pages 8–9.

Any attempt to condense the enormous subject of plant naming into a book of this size may leave some readers feeling disappointed by what has had to be omitted. However, *Plant Names Explained* is intended as a book to dip into for interest and for fun, rather than as an all-encompassing reference manual. For a scholarly, comprehensive and in-depth study of botanical names, refer to the standard work, *Botanical Latin*, by William T. Stearn (David & Charles, 1966 and 2004). A great linguist and eminent botanist, Stearn spent his life researching the naming of plants and meticulously tracing the origins of many plant names.

For up-to-date details of current plant names and availability, the recognized authority is the indispensable *RHS Plant Finder* (Dorling Kindersley Ltd), updated annually and also available on the Royal Horticultural Society's website, www.rhs.org.uk.

While every effort has been made to provide accurate and up-to-date information, plant naming is an increasingly complex business, with names changing all the time. New cultivars appear every year, while others become unavailable. The publishers therefore cannot accept responsibility for names that may have changed or been incorrectly recorded, or for plants that are no longer in commerce. We would be very happy to hear from readers should they encounter any inaccuracies.

PRIMULA VERIS

WHERE TO FIND ...

Deciphering plant names

The naming of plants is an enormous, complex and often confusing subject, but one that can be every bit as useful and entertaining to gardeners as to botanists and taxonomists. Intended for gardeners and anyone with a passing interest in plants, *Plant Names Explained* sets out to demystify botanical nomenclature and brings to life a subject that can often seem somewhat bewildering.

Why botanical Latin?

Correct botanical naming gives every plant a unique label, so that it can be accurately identified in any language: these names can be used and understood anywhere in the world. Our internationally accepted system of botanical Latin – now well over 200 years old – has become even more valuable in recent decades, as widespread travel and electronic communication have made the world ever smaller. Many of us travel widely and see unfamiliar plants in the gardens, parks and countryside of the places we visit. The nursery trade itself has become increasingly internationalized. The Internet enables us to converse across the world at the touch of a key. It has also become a very useful means of tracking down elusive plants. The system of binomial nomenclature gives us a ready-made tool to ensure that the names of the plants we see, or want to buy, or are eager to identify correctly, are immediately recognizable anywhere.

When visiting a nursery in Japan that specializes in perennials, I had no trouble identifying the plants because their names were clearly written in botanical Latin as well as in Japanese. On another occasion I was walking in a nature reserve in Slovakia with the reserve botanist. Our only method of communication was botanical Latin – accompanied by a bit of smiling and hand waving – but it was surprising how much information we managed to convey to each other.

All countries have common names for their native plants. These names are often delightfully descriptive, with many plants having a range of quirky

Bluebells and bluebells

The common name 'bluebell' shows how, in an international context, sometimes only a botanical name makes it clear which plant we mean. In England the bluebell is *Hyacinthoides non-scripta*, a bulb that makes sheets of blue in the woods in spring. In Scotland the bluebell is *Campanula rotundifolia*, a small summer flower of open grassland known in England as a harebell. In the United States the Virginia bluebell is a pretty perennial, whose showers of lilac-blue bells carpet the woods: its botanical name is *Mertensia virginica*. To Australians, a bluebell is the climber called *Sollya heterophylla*, which dangles its little sky-blue bells among its neighbours.

and charming colloquial names. However, sometimes these names can be ambiguous or obscure (see 'Bluebells and bluebells', below left). This may be because the same common name means one plant in one place but elsewhere refers to an entirely different plant, or it may be because the local name is very local indeed, perhaps known only in a dialect that is restricted to a very small area. Plants that occur in different countries or different regions may acquire a number of names, and confusion soon arises.

Carl Linnaeus

For centuries, Europe's men of science and learning communicated in Latin. Plant descriptions were written in a form of Latin, which evolved as a special language of botanical terms, taking in names from Greek, Arabic and other sources. In the 18th century, when many new plant and animal discoveries were being made, the right man appeared in the right place at the right time. Carl Linnaeus (1707–1778) was a professor of medicine at the respected University of Uppsala in Sweden. The life's work of this great naturalist was to bring order into the system of naming all living organisms. His detailed and orderly approach enabled him to build a system of classifying plants according to particular botanical details, giving each plant two names: the genus (plural: genera) and the species.

The binomial system

This binomial ('two names') system, which Linnaeus set out in his book *Species Plantarum* (1753), gained wide acceptance and was the foundation of modern botanical nomenclature. In a binomial plant name, the genus comes first (always with an initial capital letter), followed by the species name (in lower case). The names are usually written in italics: for example *Bellis perennis* (*Bellis* – genus, *perennis* – species) for the common daisy. Many genus names tell us nothing about the

Linnaeus's choice

Linnaeus often named plants for his friends – sometimes very aptly. He chose *Tillandsia* for his colleague Elias Til-Landz, who had a horror of water and once took a 2,000-mile journey overland from Finland to Sweden rather than sail across the Gulf of Bothnia. What are tillandsias? Mosses or air plants, which take their moisture from the air and require no water!

plant. More often they are adapted from another language, or were chosen to honour a particular person. The second name, the specific epithet, usually does say something about the plant. Sometimes it describes a structural detail (which may be clear only to botanists) or it indicates the plant's place of origin – either habitat or locality – or its habit of growth; thus *Bellis perennis* means perennial daisy.

The 'What's in a Name?' features in this book highlight the meanings of hundreds of descriptive terms like this. The information contained in specific epithets of plants can be very useful to those trying to grow them. For example, it's no good planting *Euphorbia palustris* in dry gravel, as *palustris* means marsh-loving. The specific epithet of *Geranium saxatile* tells us that this plant grows amongst rocks; likewise, *Geranium pratense* prefers meadows, while *Geranium sylvaticum* likes woods. Place-names can also give an indication of the geographical origin of the plant: *Geranium pyrenaicum* grows in the Pyrenees, and *Crocus corsicus* is from the Mediterranean island of Corsica. Many other epithets describe the shape of the leaves (*latifolius*, broad-leaved), the habit of growth of the plant (*humilis*, dwarf) or the colour of its flowers (*albiflorus*, white-flowered).

A further category is the specific epithet that commemorates a person, such as one of the great plant hunters. If you know the area of the world in which the plant hunter worked, you have a good idea of the plant's country, or even region, of origin. Thus many plants named for George Forrest and Reginald Farrer come from China, including the winter-flowering, scented shrub *Viburnum farreri*. The climbing *Clematis florida* var. *sieboldiana* comes from Japan, where Philipp von Siebold was collecting plants, while the epithet *douglasii*, in honour of David Douglas, usually indicates a plant that originally hailed from North America.

The gender issue

The genus name is a noun and the specific epithet a qualifying adjective. Botanical nomenclature is sourced mainly from classical Latin and Greek, where words have different endings for different genders – masculine, feminine and neuter. The specific epithet normally takes the gender of the genus name.

Latin generic names ending in -*us* are mostly masculine (as in *Helleborus foetidus*), those ending in -*a* are feminine (*Campanula alpina*), and those ending in -*um* are neuter (*Trillium grandiforum*). A notable exception is that most trees are feminine (*Prunus incisa*, *Fagus sylvatica*). Names derived from Greek are masculine if the ending is -*on* (*Platycodon grandiflorus*) and those ending in -*is* and -*ins* are feminine, as are -*ix* and -*odes* and -*oides* (*Bellis rotundifolia*, *Larix decidua*).

The most common endings for specific epithets that are Latin adjectives are: -*us* (masculine), -*a* (feminine), -*um* (neuter) – e.g. *floridus*; -*er* (masculine), -*ra* (feminine), -*rum* (neuter) – e.g. *niger*; -*is* (masculine and feminine), -*e* (neuter) – e.g. *brevis*. Certain epithets are the same in all genders, for example *simplex*, *praecox*, *tenax*, *repens*, *bicolor*, etc. Where a specific epithet denotes a person, the ending is usually -*ii* for a man (*Epimedium davidii*) and -*ae* for a woman (*Rosa banksiae*). In this book, specific epithets are listed in their masculine singular form (*angustifolius*, *niger* etc.)

Those responsible for naming plants have not always got it right. Sometimes insufficient information was available when plants were first discovered and described. For example, the shocking pink Guernsey lily, *Nerine sarniensis*, (*sarniensis* means from Guernsey) originally came from South Africa, but it is thought that the ship carrying the bulbs from the Cape was wrecked and the bulbs were washed ashore in Guernsey, where they naturalized. *Scilla peruviana* comes not from Peru, but the western Mediterrean.

In the wild, plants sometimes occur that vary slightly from the type species, perhaps in colour or habit. These are known as subspecies (subsp.). The Eastern European variation of the common primrose, which has purplish-pink flowers, is named *Primula vulgaris* subsp. *sibthorpii*. Even smaller differences are described as varieties (var.) or sometimes forms (f.). There are selections of the British native beech, *Fagus sylvatica*, that have narrower leaves than the type, with lobed and cut edges: these are known as *Fagus sylvatica* var. *heterophylla* (*heterophylla*, with diverse leaves).

Some plants are known to be hybrids between two species – either as a result of deliberate breeding, or occurring naturally where the two species grow together. Their 'hybrid epithet' is preceded by a multiplication sign, as in *Helleborus* × *nigercors*, a hybrid between *Helleborus niger* and *Helleborus argutifolius*. In the much rarer case of a hybrid between two genera (a bigeneric hybrid), a multiplication sign precedes the 'genus' name: × *Fatshedera lizei* is a bigeneric hybrid between *Fatsia japonica* and *Hedera hibernica*.

Cultivar names

The binomial system of genus and species works well for wild plants, but for cultivated plants – that is, forms of the species that have been selected and brought into cultivation, or deliberately hybridized in gardens and nurseries – a third name is necessary. This is known as the cultivar name, 'cultivar' being an abbreviation of 'cultivated variety'. The accepted way of writing a cultivar name is to place it within single quotation marks, not in italics, and with initial capital letters, as in *Bellis perennis* 'Dresden China'.

In the past, descriptive forms of Latin were used in cultivar names and many of these are still in use. However, since 1959, when the International Code of Nomenclature was established, it has not been permissible to Latinize

Plants for Amelia

It is fun to find plants that share a name with friends or members of your family. My little granddaughter is named Amelia and I was delighted to find 11 plants with 'Amelia' in their cultivar name, ranging from a pine tree called 'Amelia's Dwarf' to a rose 'Amelia Louise', and from an osteospermum called 'Sunny Amelia' to a pretty little yellow viola 'Amelia'. This seems the best one to grow in my garden.

Much fun can be had finding plants to commemorate people and special occasions: see the alphabetical 'Personal Names' panels in this book for examples of the many 'name' plants. More can be found in the *RHS Plant Finder* (see page 7).

new cultivar names. So *Bellis perennis* 'Alba Plena' for the double white form would not be acceptable as a new name nowadays.

Cultivar names come in their thousands. Quirky or informative, useful or intriguing, named for people, gardens, cities, colours or even food – a wide selection of examples features in special panels distributed throughout the book. An increasing number of cultivar names are from non-English-speaking countries, and translations of some examples of those are included here too. Knowing the meaning of a foreign name can make it easier to remember. Some excellent new clematis are being introduced from Poland and Estonia, and it may be helpful to know that *Clematis* 'Blekitny Aniol' is Polish for 'blue angel' and *Clematis* 'Sinee Dozhd' Russian for 'blue rain'. Most English-speaking gardeners also need help to decipher the many Japanese plant names. The delicate, early-flowering *Prunus incisa* 'Kojo-no-mai' becomes a 'must have' plant when you know its wonderfully evocative translation: 'dance in the ancient castle'.

Trade designations

With the introduction of Plant Breeders' Rights (PBR), cultivar names of a different kind are being registered in ever increasing numbers. Many of these are in a meaningless code that is useful in international commerce but is not suitable for marketing the plant locally.

Newer roses often bear these rather strange cultivar names – for example 'Auscat', for the rose that is usually known as Winchester Cathedral (the prefix 'Aus' indicates that the rose was bred by David Austin Roses). For everyday purposes it therefore became necessary to use a more attractive and understandable selling name, or trade designation. This name is written without quotation marks, starts with a capital letter, and is shown in a different font, as in *Rosa* WINCHESTER CATHEDRAL ('Auscat'), with the trade designation first and the cultivar name following it in brackets.

What's in a name? Three plantsmen

Tolmiea menziesii 'Taff's Gold', a North American woodland plant, is the only plant I know that carries the names of three noted plantsmen. William Fraser Tolmie was a botanist working in the Vancouver area in the 1830s. Archibald Menzies travelled as a surgeon-botanist with Captain Vancouver on his expedition to find a passage between the North Pacific and the Atlantic. Stephen Taffler, a modern plantsman with a passion for all variegated plants, was immediately attracted by the plant's gold-splashed leaves and introduced it as a garden plant.

Tolmiea menziesii 'Taff's Gold' is commonly known as the pick-a-back or piggyback plant, referring to its curious habit of producing little plantlets along its leaf veins. Related to heucheras, it is perfectly hardy but is also successful as a houseplant.

Changing names

Changes in long-established names are a nuisance for everyone and may seem difficult to justify – especially where a familiar name changes to something that is hard to remember and to spell, and even more difficult to pronounce. However, there are usually good reasons for the change. The 'principle of priority', as it is known, has established that the first written description of a plant is taken as its valid name. Familiar names can therefore sometimes fall from grace as new information comes to light from ever more obscure places around the globe. Herbaria in libraries in the former Soviet Union are now being studied by experts and we are having to accept earlier names for some plants familiar to generations of gardeners. The 'new' name may mean nothing to us and may be very difficult for many of us to pronounce – try saying *Allium przewalskianum* quickly.

Another reason for name changes is that modern scientists are able to detect ever more minute differences (and similarities) in plant characteristics. This means that sometimes plants must be separated out into a new genus: our old friend *Senecio* 'Sunshine' became *Brachyglottis* 'Sunshine', while the once-familiar genus *Cimicifuga* has disappeared into another genus, *Actaea*. As for the florists' chrysanthemum, it was despatched to a new genus called *Dendranthema*, only to be returned to *Chrysanthemum* a few years later. Occasionally taxonomists run out

Pronunciation: be bold

Pronouncing Latin names flusters many gardeners needlessly. Remember that we are quite used to saying many of them: *Aquilegia* and *Antirrhinum*, *Primula* and *Rhododendron* cause us no problems. Courage is the best principle: just have a go. Say the word aloud several times to hear how it sounds. In most cases the stress is placed on the second or third syllable. There is often no wrong or right way: ask three taxonomists, and you may well get three different versions. Never mind; we can all enjoy saying 'silly bum' for *Silybum marianum*, the milk thistle.

of names and resort to using an anagram of an existing one: the genus *Saruma* has been created from the closely related *Asarum*.

The following pages contain something for everyone. Skip from name to intriguing name, pausing to take in the riches and the oddities of the world of plant names. What plants would make a good wedding present? Who was Miss Willmott? Where does *Verbena bonariensis* come from? And how could there possibly be a plant named *Allium paradoxum* var. *normale*, or one called *Quisqualis indica*? Dip into *Plant Names Explained* for the answers to hundreds of questions like these. You will find it useful, absorbing and fun.

Jane Sterndale-Bennett

A

a- without (e.g. *acaulis,* stemless)
abbreviatus shortened
abies; abietinus fir; resembling a fir
  Picea abies
abnormis abnormal
  Ranunculus abnormis
abrotanifolius with leaves
  like *Artemisia abrotanum*
  (southernwood)
  Geranium abrotanifolium
abscissus cut off
absconditus hidden
absinthoides like *Artemisia*
  absinthium (wormwood)
  Erodium absinthoides
abyssinicus from Ethiopia (formerly
  Abyssinia)
acaciiformis shaped like *Acacia*
  Eucalyptus acaciiformis
acadiensis from Nova Scotia,
  Canada
acalycinus without (or apparently
  without) a calyx
  Liquidambar acalycina
acanth-, acantho- spiny, thorny
  Onopordum acanthium
acanthifolius with leaves like
  Acanthus
  Carlina acanthifolia
acanthos a prickle or thorn

COMMON DESCRIPTIVE TERMS
acaulis, acaulescens stemless
Carlina acaulis
Draba acaulis
Gentiana acaulis
Oenothera acaulis
Silene acaulis

acephalus without a head
acer, acris, acre sharp, pointed,
  piercing
  Sedum acre
acerbus bitter
acerifolius with leaves like *Acer,*
  maple-leaved
  Viburnum acerifolium
aceroides like *Acer* (maple)
acerosus needle-shaped
acerus without wax
acetabulosus cup-shaped,
  concave
  Ballota acetabulosa

CARLINA ACAULIS

acetosus, acetosellus sour, acid
Oxalis acetosella
Pelargonium acetosum
acicularis like a pin or needle
Eleocharis acicularis
acidus sour, acid
aciformis needle-shaped
acinaceus like a scimitar
acinosus like grapes
Phytolacca acinosa
acis something pointed
aconitifolius with leaves like
Aconitum
Ampelopsis aconitifolia
acraeus living on the heights
Euryops acraeus
acris, acre see *acer*
acro- the tip or extremity
of something
Rhododendron acrophilum

acta the seashore
Festuca actae
actino- radiating (see below)
actinophyllus with radiating leaves
Schefflera actinophylla
actinosus full of rays; glorious
aculeatus prickly
Ruscus aculeatus

COMMON DESCRIPTIVE TERMS
acuminatus tapering to a narrow
point
Gladiolus acuminatus
Lonicera acuminata
Magnolia acuminata
Rheum acuminatum
Tulipa acuminata

---

acus a needle or pin
acutatus sharpened
Helichrysum acutatum
acutiflorus with sharply pointed
flowers
Calamagrostis × acutiflora
acutifolius with sharply pointed
leaves
Salix acutifolia
acutiformis in the shape of a point
Carex acutiformis
acutissimus very sharp
Quercus acutissima
acutus sharpened, made pointed
Carex acuta
adamantinus hard like steel

Aaron's beard *Hypericum calycinum*
Aaron's rod *Solidago*; *Verbascum thapsus*
Abele *Populus alba*
Aconite *Aconitum* or *Eranthis*
Adam's needle *Yucca filamentosa*
Adder's tongue *Erythronium americanum*
African daisy *Arctotis*
African lily *Agapanthus*
African marigold *Tagetes erecta*
Alder *Alnus*
Alder buckthorn *Rhamnus frangula*
Alexanders *Smyrnium olusatrum*
Alexandrian laurel *Danae racemosa*
Alfalfa *Medicago sativa*
Alkanet *Anchusa*; *Alkanna tinctoria*; *Pentaglottis sempervirens*
Allspice *Calycanthus*
Almond *Prunus dulcis*
Alpenrose *Rhododendron ferrugineum*

Alumroot *Heuchera*
Alyssum, Sweet *Lobularia maritima*
Angel's fishing rod *Dierama pulcherrimum*
Angel's tears *Narcissus triandrus*
Angel's trumpets *Brugmansia*
Angelica *Angelica archangelica*
Angelica tree *Aralia spinosa*
Apple *Malus domestica*
Apple of Peru *Nicandra physalodes*
Apricot *Prunus armeniaca*
Arborvitae *Thuja*
Arum lily *Zantedeschia aethiopica*
Ash *Fraxinus*
Aspen *Populus tremula*
Aster (China) *Callistephus*
Atlas cedar *Cedrus atlantica*
Autograph tree *Clusia major*
Avens *Geum*
Azalea *Rhododendron*

adeno- gland
adenocarpus with glandular (or sticky) fruits
Helichrysum adenocarpum
adenogynus with a sticky ovary
Rhododendron adenogynum
adenophorus bearing glands
Adenophora bulleyana
adenophyllus with glands on the leaves, or with sticky leaves
Oxalis adenophylla
adiantoides like *Adiantum* (maidenhair fern)
admirabilis admirable, wonderful
adnatus growing together

adpressus lying flat, fitting close to
Cotoneaster adpressus
adscendens ascending
adsurgens rising, standing up
Phlox adsurgens
aegypticus, aegyptiacus Egyptian
Hypericum aegypticum
aemulus emulating, comparable to
Dryopteris aemula
aeneus copper or bronze
aequalis equal
Phygelius aequalis
aequiformis uniform
aequinoctialis equinoctial; from equatorial (tropical) regions

aequipetalus with equal petals

aequitrilobus with three equal lobes

aerius aerial

aeruginosus blue-green (the colour of verdigris)

  Rhododendron campanulatum subsp. *aeruginosum*

aesculifolius with leaves like horse-chestnut

  Rodgersia aesculifolia

aestivalis, aestivus of summer

  Adonis aestivalis

  Leucojum aestivum

aethiopicus, aethiopis Ethiopian; also used to mean South African

  Zantedeschia aethiopica

aetnensis from Mount Etna, Sicily

  Genista aetnensis

aetolicus from Aetolia, Greece

COMMON DESCRIPTIVE TERMS

affinis related to

Dryopteris affinis

Fritillaria affinis

Gentiana affinis

Jasminum officinale f. *affine*

Persicaria affinis

Pulmonaria affinis

Tricyrtis affinis

---

affixus attached to

afghanicus from Afghanistan

  Dionysia afghanica

afoliatus without leaves

CULTIVAR NAMES
Food and Drink

ASTILBE 'LOLLIPOP'

CAMELLIA HIEMALIS 'SPARKLING BURGUNDY'

CARYOPTERIS × *CLANDONENSIS* 'SUMMER SORBET'

DAHLIA 'CAFÉ AU LAIT'

HOSTA 'GUACAMOLE'

HEUCHERA 'PLUM PUDDING'

IRIS SIBIRICA 'BUTTER AND SUGAR'

IRIS 'SMOKED SALMON'

MALUS DOMESTICA (APPLE) 'SCOTCH DUMPLING'

MENTHA ARVENSIS 'BANANA'

PAPAVER ORIENTALE 'WATERMELON'

PELARGONIUM 'TURKISH COFFEE'

PENSTEMON 'SOUR GRAPES'

POTENTILLA FRUTICOSA 'PEACHES AND CREAM'

PRIMULA AURICULA 'OLD MUSTARD'

PULMONARIA 'SPILLED MILK'

RANUNCULUS REPENS 'BUTTERED POPCORN'

RHODODENDRON SARGENTIANUM 'WHITEBAIT'

ROSA GINGER SYLLABUB ('HARJOLLY')

RUDBECKIA HIRTA 'MARMALADE'

SALVIA MERJAMIE 'MINT-SAUCE'

SAXIFRAGA 'BLACKBERRY AND APPLE PIE'

SCABIOSA ATROPURPUREA 'CHILLI SAUCE'

SEDUM 'STEWED RHUBARB MOUNTAIN'

SEMPERVIVUM 'JELLY BEAN'

SISYRINCHIUM 'RASPBERRY'

CULTIVAR NAMES
Colours: Red

ASTILBE CHINENSIS 'VISION IN RED'

ATHYRIUM NIPONICUM VAR. *PICTUM*
'RED BEAUTY'

BEGONIA 'RED UNDIES'

BERBERIS THUNBERGII F. *ATROPURPUREA*
'RED CHIEF'

BUDDLEJA DAVIDII 'ROYAL RED'

CANNA INDICA 'RUSSIAN RED'

CORYLUS MAXIMA 'RED FILBERT'

ENKIANTHUS CAMPANULATUS
'RED BELLS'

ESCALLONIA 'RED ELF'

EUONYMUS EUROPAEUS
'RED CASCADE'

HEBE 'RED EDGE'

HELLEBORUS × *HYBRIDUS* 'RED LADY'

HEMEROCALLIS 'LITTLE RED HEN'

ILEX VERTICILLATA 'WINTER RED'

IRIS SIBIRICA 'MELTON RED FLARE'

LATHYRUS LATIFOLIUS 'RED PEARL'

LAVANDULA STOECHAS 'KEW RED'

LEPTOSPERMUM SCOPARIUM
'RED DAMASK'

LONICERA PERICLYMENUM
'RED GABLES'

MALUS DOMESTICA (APPLE)
'RED FALSTAFF'

MALUS × *ROBUSTA* 'RED SENTINEL'

NYMPHAEA 'PERRY'S BABY RED'

PELARGONIUM
'RED BLACK VESUVIUS'

PERSICARIA MICROCEPHALA
'RED DRAGON'

PHOTINIA × *FRASERI* 'RED ROBIN'

SEMPERVIVUM 'RED SPIDER'

COMMON DESCRIPTIVE TERMS

africanus; afer, afra, afrum
African, from Africa

Agapanthus africanus

Artemisia afra

Cyclamen africanum

Salvia africana

---

agathos good

agathosmus smelling good, fragrant
Buddleja agathosma

agavifolius with leaves like *Agave*
Eryngium agavifolium

agavoides like *Agave*
Echeveria agavoides

ageratifolius with leaves like *Ageratum*
Achillea ageratifolia

ageratoides like *Ageratum*

ageraton not growing old; not
withering readily
Ageratum houstonianum

agglomeratus gathered into a head

aggregatus added or collected
together; in a cluster
Eucalyptus aggregata

agrarius of land or fields

agrestis rural, rustic; of fields

agrius wild

agrostis grass

ailanthifolius with leaves like
Ailanthus
Juglans ailanthifolia

aizoon ever-living
Sedum aizoon

akebioides resembling *Akebia*
   Clematis akebioides
alaris on the wing; axillary

---

alatus winged
Euonymus alatus
Gladiolus alatus
Nicotiana alata
Passiflora alata
Thunbergia alata

---

albanus, albanicus from Albania
   Geranium albanum
albescens becoming white, whitish
albicans whitish
   Hebe albicans
albicaulis white-stemmed
albidus white
   Nymphaea 'Albida'
albiflorus, albiflos white-flowered
   Geranium albiflorum
albifrons with white foliage or fronds
   Lupinus albifrons
albispinus with white thorns
albocereus white-waxy
   Fargesia albocerea
albomaculatus white-spotted
   Zantedeschia albomaculata
albopictus white-painted
   Begonia albopicta
albopilosus with white hairs
albosinensis Chinese white
   Betula albosinensis

albospicus with white spikes
albomarginatus white-edged
albotomentosus white-woolly
   Stachys albotomentosa
albovariegatus white-variegated
albulus whitish
   Carex albula

---

albus white
Cornus alba
Dictamnus albus
Nymphaea alba
Phlox paniculata 'Alba Grandiflora'
Populus alba
Rosa 'Alba Semiplena'
Veratrum album

VERATRUM ALBUM

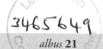

CULTIVAR NAMES
Personal Names
A

KNIPHOFIA 'ADA'	*PELARGONIUM* 'ANGELA'
ROSA 'ADAM'	*TULIPA* 'ANGÉLIQUE'
PRIMULA AURICULA 'ADRIAN'	*PRIMULA AURICULA* 'ANGIE'
ROSA × *FRANCOFURTANA* 'AGATHA'	*CLEMATIS* 'ANITA'
ROSA 'AGNES'	*SALVIA* × *SUPERBA* 'DEAR ANJA'
HEMEROCALLIS 'ALAN'	*MAGNOLIA* 'ANN'
ROSA ALEC'S RED ('CORED')	*CLEMATIS* ANNA LOUISE ('EVITHREE')
PELARGONIUM 'ALEX'	*HEDERA HELIX* 'ANNA MARIE'
RHODODENDRON 'ALEXANDER'	*HELIANTHEMUM* 'ANNABEL'
CALLUNA VULGARIS 'ALEXANDRA'	*IRIS* 'ANNABEL JANE'
PRIMULA AURICULA 'ALF'	*RHODODENDRON* 'ANNABELLA'
RHODODENDRON 'ALFRED'	*HYDRANGEA ARBORESCENS* 'ANNABELLE'
DIANTHUS 'ALICE'	*VERBENA* 'SILVER ANNE'
PRIMULA AURICULA 'ALICIA'	*CALLUNA VULGARIS* 'SISTER ANNE'
ROSA ALISON ('COCLIBEE')	*CALLUNA VULGARIS* 'ANNEMARIE'
CROCOSMIA 'ALISTAIR'	*POTENTILLA FRUTICOSA* 'ANNETTE'
SEMPERVIVUM 'AMANDA'	*VERBASCUM* 'ANNIE MAY'
VIOLA 'AMELIA'	*ROSA* 'ANTONIA'
NEMESIA 'AMÉLIE'	*RHODODENDRON* 'ANTONIO'
HEBE 'AMY'	*VIOLA* 'ARABELLA'
ROSA ANABELL ('KORBELL')	*PRIMULA AURICULA* 'UNCLE ARTHUR'
CHRYSANTHEMUM 'ANASTASIA'	*ASTER NOVI-BELGII* 'AUDREY'
ROSA 'ANDREA'	*PELARGONIUM* 'AUGUSTA'
FUCHSIA 'ANDREW'	*PRIMULA AURICULA* 'AVRIL'

aleppensis, aleppicus from Aleppo, Syria
   Geum aleppicum
aleuticus from the Aleutian Islands
   Adiantum aleuticum
alexandrinus from Alexandria, Egypt
algeriensis from Algeria
   Salvia algeriensis

algidus cold
   Primula algida
alienus alien, foreign
   Quercus aliena
aliformis wing-shaped
alleghaniensis from the Allegheny Mountains in the eastern USA
   Betula alleghaniensis

alliaceus like *Allium* (onion)
allo- diverse, different, other
alnifolius with leaves like *Alnus*
(alder)
  Clethra alnifolia
alnoides resembling *Alnus* (alder)
  Betula alnoides
aloides like Aloe
aloifolius with leaves like *Aloe*
  Yucca aloifolia
alpestris growing just below the
alpine zone
  Ranunculus alpestris
alpicola an inhabitant of the
alpine zone
  Primula alpicola

---

alpinus alpine, growing in the Alps
or in the alpine zone
Aquilegia alpina
Clematis alpina
Dianthus alpinus
Eryngium alpinum
Lychnis alpina
Papaver alpinum

---

altaclerensis from Highclere in
Hampshire
  Ilex × *altaclerensis*
altaicus from the Altai Mountains
in southern Siberia/Kazakhstan
  Primula altaica
alternatus alternate

alternifolius with alternate leaves
  Buddleja alternifolia
althaeoides like *Althaea* (hollyhock)
  Convolvulus althaeoides
alticola hill dweller
alticomus with foliage high up
altifrons with tall foliage or fronds
altipendulus hanging high
altissimus very tall
  Ailanthus altissima
altus tall
  Sedum altum
alutaceus leathery
  Rhododendron alutaceum
alveatus, alveolatus hollowed out,
channelled
alyssifolius with leaves like *Alyssum*
  Phlox alyssifolia
alyssoides like *Alyssum*
  Halimium alyssoides
amabilis lovely
  Kolkwitzia amabilis
amanus from the Amanus mountain
range in Turkey
  Origanum amanum
amaranthoides like *Amaranthus*
amarellus rather bitter
amarissimus very bitter
  Salvia amarissima
amarus bitter
amarylloides like *Amaryllis*
  Haemanthus amarylloides
amazonicus from the Amazon
  Eucharis amazonica

CULTIVAR NAMES
Stage and Screen
RHODODENDRON 'BAMBI'

ROSA INGRID BERGMAN 'POULMAN'

PAEONIA LACTIFLORA 'SARAH BERNHARDT'

RHODODENDRON 'BLUE PETER'

DAHLIA 'DORIS DAY'

PRIMULA AURICULA 'MARGOT FONTEYN'

FUCHSIA 'AUDREY HEPBURN'

PAEONIA 'HIGH NOON'

CAMELLIA 'BOB HOPE'

SEMPERVIVUM 'KERMIT'

NARCISSUS 'CAROLE LOMBARD'

RHODODENDRON 'MARY POPPINS'

PELARGONIUM 'PYGMALION'

PELARGONIUM 'BERYL REID'

FUCHSIA 'ANGELA RIPPON'

PELARGONIUM 'GINGER ROGERS'

DELPHINIUM 'SIR HARRY SECOMBE'

PAEONIA LACTIFLORA 'SHIRLEY TEMPLE'

NARCISSUS 'SPENCER TRACY'

PRIMULA AURICULA 'JOHN WAYNE'

FUCHSIA 'BARBARA WINDSOR'

COMMON DESCRIPTIVE TERMS

americanus American

Agave americana

Erythronium americanum

Fraxinus americana

Lysichiton americanus

Sorbus americana

Tilia americana

---

amethystinus violet-coloured

Festuca amethystina

ammophilus sand-loving

amoenus lovely, pleasant

Felicia amoena

amomum an aromatic shrub

Cornus amomum

amphi- both

amphibius amphibious, growing in water and on land

amplexicaulis clasping or encircling the stem

Persicaria amplexicaule

amplexifolius with leaves that clasp or encircle the stem

---

ambiguus doubtful or uncertain

Consolida ambigua

ambly- blunt

amblyanthus with blunt flowers

Indigofera amblyantha

amblyphyllus blunt-leaved

ambrosioides like ambrosia, the food of the gods

amentaceus like a catkin (*amentum*)

COMMON DESCRIPTIVE TERMS

amurensis from the area of Heilong Jiang (the Amur river), on the Russian/Chinese border

Adonis amurensis

Dianthus amurensis

Maackia amurensis

Papaver amurense

Vitis amurensis

amygdaleus of an almond tree
amygdalinus, amygdaloides
  almond-like
  Eucalyptus amygdalina
  Euphorbia amygdaloides
amylaceus starchy
an- without
ana- up
anatolicus from Anatolia, Turkey
  Dianthus anatolicus
anceps with two edges
  Lythrum anceps
ancyrensis from Ankara, Turkey
  Crocus ancyrensis
andinus from the Andes
  Prumnopitys andina
androgynus hermaphrodite
  Semele androgyna
anemoniflorus anemone-flowered
  Paeonia officinalis 'Anemoniflora
  Rosea'
aneurus without nerves
anglicus English
  Sorbus anglica
anguilliformis eel-shaped
anguinus snakelike
angularis, angulatus angular, with
  angles
  Geranium sylvaticum
  'Angulatum'
angulosus with angles or corners
  Allium angulosum
angustatus narrowed
  Cornus angustata

COMMON DESCRIPTIVE TERMS
angustifolius narrow-leaved
Elaeagnus angustifolia
Fraxinus angustifolia
Galanthus angustifolius
Hoheria angustifolia
Lavandula angustifolia
Phillyrea angustifolia
Pulmonaria angustifolia
Typha angustifolia

LEAF CHARACTERISTICS

acutifolius with sharply pointed leaves

adenophyllus with glands on the leaves, or with sticky leaves

alternifolius with alternate leaves

angustifolius narrow-leaved

aquifolius with pointed leaves

argophyllus silver-leaved

argutifolius with sharply toothed leaves

brevifolius with short leaves

capillifolius with hairlike leaves

cardiophyllus with heart-shaped leaves

caudatifolius with tail-like leaves

chrysophyllus with golden leaves

cochlearifolius with spoon-shaped leaves

cordifolius with heart-shaped leaves

crassifolius with thick leaves

dactylifolius with finger-like leaves

dentatus toothed

denticulatus with small teeth

dictyophyllus with leaves showing an obvious network of veins

digitatus hand-shaped; with fingers

dissectus dissected; deeply divided

distichophyllus with leaves in two rows

diversifolius with leaves of differing shapes

filicifolius with fern-like leaves

filifolius thread-leaved

glaucophyllus with glaucous leaves

grandidentatus with large teeth

grandifolius with large leaves

integrifolius with undivided leaves

isophyllus with equal leaves

latifolius broad-leaved

laxifolius loose-leaved

linearifolius with linear leaves

lobularis, lobulatus with small lobes

longifolius long-leaved

macrophyllus with large leaves

megalophyllus with large leaves

microphyllus with small leaves

millefolius, millefoliatus (literally) with a thousand leaves

mucronifolius with pointed leaves

oblongifolius with oblong leaves

obtusifolius with blunt leaves

oppositifolius opposite-leaved

ovalifolius with oval leaves

palmatus palmate

parvifolius with small leaves

pectinatus like a comb, combed

perfoliatus perfoliate (with the leaf surrounding the stem)

petiolaris with a (long) leaf stalk

pinguifolius with fat leaves

pinnatus pinnate

planifolius with flat leaves

platyphyllus, platyphyllos with broad (or flat) leaves

plicatus folded or pleated

podophyllus with stalked leaves

polifolius grey-leaved

porphyrophyllus with purple leaves

repandens, repandus with wavy or turned-up edges

rhombifolius with diamond-shaped leaves

rotundifolius with round leaves

rubrifolius with red leaves

sagittifolius with arrow-shaped leaves

salicifolius with willow-like leaves

serratifolius with serrated leaves

serratus serrated, saw-toothed

sessilifolius with sessile leaves

stenophyllus with narrow leaves

tenuifolius with thin leaves

HELIANTHUS ANNUUS

angustipetalus narrow-petalled
   Hydrangea angustipetalus
angustus narrow
   Gladiolus angustus
anisatus, anisus aniseed-flavoured
   Illicium anisatum
   Pimpinella anisum
annularis ring-shaped
annulatus ringed; with rings
annuus annual
   Helianthus annuus (sunflower)
anomalus irregular, abnormal
   Hydrangea anomala subsp. *petiolaris*
anopetalus with upright petals
anserinus pertaining to geese
   Potentilla anserina
antarcticus from Antarctica
   Dicksonia antarctica

anthemoides like *Anthemis*
   Rhodanthe anthemoides
anthemon flower
anthera anther (the pollen-bearing
   part of a flower's stamen)
anthopogon flower beard (hairs or
   bristles)
   Rhododendron anthopogon
anthracinus coal black
-anthus -flowered (e.g. *macranthus*,
   large-flowered)
antillaris from the Antilles (West
   Indies)
antipodeus from the Antipodes
   Geranium × *antipodeum*
antiquus old, ancient, former
apenninus from the Apennines, Italy
   Anemone apennina

ILEX AQUIFOLIUM

apertus open
  Nomocharis aperta
apetalus without petals
  Fuchsia apetala
aphyllus without leaves
apianus of bees
  Salvia apiana
apiculatus with a short pointed
  end
  Luma apiculata
apifer bee-bearing
  Ophrys apifera (bee orchid)
apiifolius with leaves like parsley
  or celery
  Clematis apiifolia

apo- down
apodus without a foot or stalk;
  sessile
  Salix apoda
appendiculatus with small
  appendages
  Francoa appendiculata
applanatus flattened
applicatus joined, attached
appressus see *adpressus*
approximans approaching, near to
  Eucalyptus approximans
apterus without wings
  Euonymus alatus var. *apterus*
aquaticus, aquatilis of water; growing
  by water
  Mentha aquatica
  Ranunculus aquatilis
aqueus watery
aquifolius with pointed leaves
  Ilex aquifolium (holly)
aquilegiifolius with leaves like
  Aquilegia
  Thalictrum aquilegiifolium
aquilinus eagle-like
arabicus Arabian
  Ornithogalum arabicum
arachnoides, arachnoideus like
  a spider or its web
  Sempervivum arachnoideum
aragonensis from Aragon, Spain
arborescens growing to be a tree;
  woody
  Artemisia arborescens

arboreus treelike, woody
Brugmansia arborea
Ceanothus arboreus
Erica arborea
Lavatera arborea
Linum arboreum
Lupinus arboreus
Pseudopanax arboreus
Rhododendron arboreum

arbusculus like a small tree
  Daphne arbuscula
arbutifolius with leaves like *Arbutus*
  Aronia arbutifolia
arcanus closed, secret
  Phyllostachys arcana
archangelicus of an archangel
  Angelica archangelica
arcticus arctic
  Rubus arcticus
arcuatus bent or curved like a bow
  Ornithogalum arcuatum

arenarius, arenosus sandy, growing
  on sand
Dianthus arenarius
Jovibarba arenaria
Leymus arenarius
Sisyrinchium arenarium

areolatus marked out into small areas
  Sarracenia × *areolata*

CULTIVAR NAMES
Foreign Expressions
A

ABENDGLUT *(German)*
EVENING GLOW
(*BERGENIA* 'ABENDGLUT')

ALBÂTRE *(French)* ALABASTER
(*PHILADELPHUS* 'ALBÂTRE')

ALLEGRO *(Italian)* CHEERFUL
(*PAPAVER ORIENTALE* 'ALLEGRO')

AMANOGAWA *(Japanese)*
THE MILKY WAY
(*PRUNUS* 'AMANOGAWA')

ANDENKEN AN (…) *(German)*
REMEMBRANCE OF (…)
(*PENSTEMON* 'ANDENKEN AN
FRIEDRICH HAHN')

APFELBLÜTE *(German)*
APPLE BLOSSOM
(*GERANIUM SANGUINEUM*
'APFELBLÜTE')

APRÈS MOI *(French)* AFTER ME
(*HEMEROCALLIS* 'APRÈS MOI')

ARC-EN-CIEL *(French)* RAINBOW
(*NYMPHAEA* 'ARC-EN-CIEL')

ASCHERMITTWOCH *(German)*
ASH WEDNESDAY
(*ROSA* 'ASCHERMITTWOCH')

AUBADE *(French)* DAWN MUSIC
(*CROCUS CHRYSANTHUS* 'AUBADE')

AZURFEE *(German)* AZURE FAIRY
(*ERIGERON* 'AZURFEE')

AZUMA-KAGAMI *(Japanese)*
MIRROR OF THE EAST
(*IRIS ENSATA* 'AZUMA-KAGAMI')

argenteoguttatus silver-spotted
argenteomarginatus silver-edged
argenteovariegatus silver-variegated
   Vinca minor 'Argenteovariegata'

COMMON DESCRIPTIVE TERMS
argenteus silver, **argentatus**
   silvered
Celosia argentea
Cistus × *argenteus*
Ilex aquifolium 'Argentea Marginata'
Plectranthus argentatus
Potentilla argentea
Salvia argentea
Tanacetum argenteum

argentinus from Argentina
argillosus rich in clay, full of clay
argophyllus silver-leaved
   Olearia argophylla
argutifolius with sharply toothed
   leaves
   Helleborus argutifolius
argutus sharp
   Spiraea 'Arguta'
argyro- silvery (see below)
argyrophyllus silver-leaved
   Rhododendron argyrophyllum
argyrotrichon silver-haired
   Campanula argyrotricha
aridus dry, withered
   Penstemon aridus
arifolius with leaves like *Arum*
   Asarum arifolium

COMMON DESCRIPTIVE TERMS
aristatus, aristosus awned (like ears
   of corn); bearded
Aloe aristata
Gaillardia aristata
Pinus aristata
Thamnocalamus aristatus

arizonicus from Arizona, USA
   Abies lasiocarpa var. *arizonica*
arkansanus from Arkansas, USA
armandii named after Abbé Jean
   Pierre Armand David (1826–
   1900); see also page 65 (*davidii*)
   Clematis armandii
armatus equipped, armoured (e.g.
   with thorns)
   Osmanthus armatus
armeniacus, armenus Armenian
   Prunus armeniaca (apricot)
armillaris encircled
aromaticus fragrant, aromatic, spicy
   Clematis × *aromatica*
articulatus jointed; distinct
   Oxalis articulata
arundinaceus like a reed
   Phalaris arundinacea

COMMON DESCRIPTIVE TERMS
arvensis of fields or farmland
Cirsium arvense
Knautia arvensis
Myosotis arvensis
Rosa arvensis

arvoniensis from Caernarvon, Wales
ascendens ascending, rising
asiaticus Asian
   Trachelospermum asiaticum
asperifolius with rough leaves
asperatus roughened
   Arisaema asperatum
asper rough
   Hydrangea aspera
asperrimus very rough
aspleniifolius with leaves like the fern
   Asplenium
   Fagus sylvatica var. *heterophylla*
   'Aspleniifolia'
assurgens standing up, rising
assyriaca from Assyria
   Fritillaria assyriaca
astictus without spots
asturiensis from Asturias, in north-
   western Spain
   Narcissus asturiensis

COMMON DESCRIPTIVE TERMS
ater-, atro- dark, black
atrocyaneus very dark blue:
   Salvia atrocyanea
atropurpureus dark purple:
   Scabiosa atropurpurea 'Ace of Spades'
atrorubens dark red:
   Helleborus atrorubens
atrosanguineus dark blood-red:
   Cosmos atrosanguineus
atroviolaceus dark violet:
   Gladiolus atroviolaceus

---

atlanticus from the Atlas Mountains
   in North Africa
   Cedrus atlantica (Atlas cedar)
atratus dressed in mourning; dark
   Aquilegia atrata
atriplicifolius with leaves like *Atriplex*
   (orache)
   Perovskia atriplicifolia
atro- (see panel, below left)
attenuatus tapering to a point
   Agave attenuata
atticus from Attica or Athens,
   Greece
   Iris attica

GENUS NAMES
A

Abelia named for Dr Clarke Abel
   (1780–1826), British botanist
Acanthus from Greek *akantha*, thorn
Achillea named for the Greek hero
   Achilles
Agapanthus from Greek *agape*, love,
   and *anthos*, flower
Alstroemeria named for Baron Claus
   Alstroemer (1736–1794)
Ampelopsis from Greek *ampelos*, vine,
   and *opsis*, resemblance
Anemone from Greek *anemos*, wind
Antirrhinum from Greek *anti*, like,
   and *rhis*, a nose or snout
Artemisia named for the Greek
   goddess Artemis
Aubrieta named for Claude
   Aubriet (1668–1743), French
   botanical artist

PRIMULA AURICULA

aucuparius bird-catching
  Sorbus aucuparia
augurius of the augur or soothsayer;
  predicting
augustus venerable, majestic
aulos flute, tube
aurantiacus, aurantius orange-red
  Citrus aurantium (orange tree)
  Pilosella aurantiaca
auratus, aureatus decorated with gold
aureomaculatus gold-spotted
aureomarginatus gold-edged
  Hosta 'Aureomarginata'
aureopictus (literally) gold-painted
aureoreticulatus gold-veined
  Lonicera japonica 'Aureoreticulata'

aureovariegatus gold-variegated
  Buxus sempervirens 'Aureovariegata'

COMMON DESCRIPTIVE TERMS
aureus gold, golden, of gold
Bidens aurea
Fritillaria aurea
Humulus lupulus 'Aureus'
Juniperus chinensis 'Aurea'
Origanum vulgare 'Aureum'
Philadelphus coronarius 'Aureus'
Phyllostachys aurea
Sambucus racemosa 'Plumosa Aurea'

aureolus golden, gilded
auricomus golden-haired
  Pleioblastus auricomus
auricula ear, ear-shaped appendage
  Primula auricula
auriculatus eared, lobed
  Scrophularia auriculata
auritus eared
  Salix aurita
aurorius of dawn, daybreak, morning
aurosus like gold
australiensis from Australia

COMMON DESCRIPTIVE TERMS
australis southern; from the
  southern hemisphere
Baptisia australis
Cordyline australis
Indigofera australis
Phragmites australis

austriacus from Austria
  Veronica austriaca

COMMON DESCRIPTIVE TERMS

autumnalis of autumn
Colchicum autumnale
Helenium autumnale
Leucojum autumnale
Prunus subhirtella 'Autumnalis'
Scilla autumnalis

avellanus from Abella (now called
  Avellino), Campania, Italy (a
  town once noted for its fruit
  and nuts)
  Corylus avellana
avicularis pertaining to small birds
  Solanum aviculare
avium of birds
  Prunus avium (bird cherry; sweet
  cherry)
axillaris axillary; relating to the
  axil
  Pachysandra axillaris
azoricus from the Azores
  Hedera azorica

COMMON DESCRIPTIVE TERMS

azureus azure; deep sky-blue
Anchusa azurea
Muscari azureum
Salvia azurea
Symphytum azureum
Tropaeolum azureum

CULTIVAR NAMES
Colours: Purple

ACER PLATANOIDES
'GOLDSWORTH PURPLE'
AUBRIETA 'GREENCOURT PURPLE'
BUDDLEJA DAVIDII 'NANHO PURPLE'
CLEMATIS 'PURPLE SPIDER'
CLEMATIS HERACLEIFOLIA
'CHINA PURPLE'
CORDYLINE AUSTRALIS
'PURPLE TOWER'
CORYLOPSIS SINENSIS VAR. SINENSIS
'SPRING PURPLE'
COTINUS COGGYGRIA 'ROYAL PURPLE'
CROCUS TOMMASINIANUS
'WHITEWELL PURPLE'
ERYTHRONIUM DENS-CANIS
'PURPLE KING'
FAGUS SYLVATICA 'PURPLE FOUNTAIN'
FUCHSIA 'DEEP PURPLE'
GERANIUM PRATENSE 'PURPLE HERON'
HEMEROCALLIS 'PURPLE RAIN'
HEUCHERA MICRANTHA
'PALACE PURPLE'
HOSTA 'PURPLE DWARF'
LAVANDULA ANGUSTIFOLIA
'TWICKEL PURPLE'
LIRIOPE MUSCARI 'ROYAL PURPLE'
LYTHRUM VIRGATUM
'DROPMORE PURPLE'
OCIMUM BASILICUM VAR.
PURPURASCENS 'PURPLE RUFFLES'
PASSIFLORA 'PURPLE HAZE'
PENSTEMON 'PURPLE BEDDER'
PRUNUS PADUS 'PURPLE QUEEN'
PULMONARIA 'MOURNFUL PURPLE'
SALVIA 'PURPLE MAJESTY'

B

babylonicus Babylonian
  Salix babylonica
baccans berrying
baccatus berry-like; with pulpy fruits
  Taxus baccata
baccifer berry-bearing
bacciflavus with yellow berries
  Ilex aquifolium 'Bacciflava'
bacillaris like a rod or stick
backhouseanus named for an English
  nurseryman and plant collector,
  James Backhouse (1794–1869)
Correa backhouseana
badius reddish brown
baeticus from Andalucia (the Roman
  province of Baetica) in Spain

balcanicus from the Balkans
baldensis from Monte Baldo,
  in northern Italy
baldschuanicus from Baljuan,
  Turkestan
  Fallopia baldschuanica

COMMON DESCRIPTIVE TERMS
balearicus from the Balearic Islands,
  in the Mediterranean Sea
Arenaria balearica
Buxus balearica
Clematis cirrhosa var. *balearica*
Hypericum balearicum

---

ballardiae named for the English
  gardener and nurserywoman Helen
  Ballard (1908–1995), known
  especially for breeding hellebores
  Helleborus × *ballardiae*

**SIR JOSEPH BANKS
(1743–1820)**

Wealthy Oxford graduate and passionate naturalist Joseph Banks accompanied Captain Cook on his 1768–71 world voyage – an auspicious start to an illustrious career in the world of plants. Shortly after his return, Banks took up the post of Adviser to the Royal Gardens at Kew, presiding over the introduction of numerous new plants and helping to establish Kew's international reputation. He was President of the Royal Society from 1778 until his death. Banks always took a particular interest in the plant life of Australia and it is fitting that he gave his name to an Australasian genus – *Banksia*. Other plants commemorating him include *Astelia banksii* and *Pinus banksiana*, while *Rosa banksiae* is named for his wife.

balsameus balsamic or resinous
  Abies balsamea
balsamifer balsam-bearing
  Populus balsamifera
balteiformis shaped like a belt
balticus from the Baltic Sea region
bambusoides like bamboo
  Phyllostachys bambusoides
banaticus, bannaticus from Banat,
  Romania
  Echinops bannaticus
banksii, banksiae (see panel opposite)
barbadensis from Barbados in the
  West Indies
barba-jovis Jove's beard
  Anthyllis barba-jovis
barbarus foreign; from the Barbary
  Coast of Africa
  Decumaria barbara
barbatus bearded
  Dianthus barbatus (sweet
  William)
barbinervis with hairs on the veins
  Clethra barbinervis
barbulatus with a small beard
  Anemone barbulatus
barystachys with heavy spikes
  Lysimachia barystachys
basilicus princely, royal
  Ocimum basilicum (sweet basil)
battandieri named for Jules Battandier
  (1848–1922), a French botanist
  Cytisus battandieri
bavaricus from Bavaria

CULTIVAR NAMES
Personal Names
B

GALANTHUS 'BARBARA'S DOUBLE'
HOSTA 'BARBARA ANN'
DAHLIA 'BEATRICE'
PELARGONIUM 'BEATRIX'
LEUCANTHEMUM × *SUPERBUM* 'BECKY'
FUCHSIA 'BELINDA JANE'
CLEMATIS 'BELLA'
CAMELLIA SASANQUA 'BEN'
CAMPANULA TRACHELIUM 'BERNICE'
NARCISSUS 'BERYL'
PYRUS COMMUNIS (PEAR) 'BETH'
MAGNOLIA 'BETTY'
FUCHSIA 'BEVERLEY'
FUCHSIA 'BRENDA'
ESCALLONIA LAEVIS 'GOLD BRIAN'
HEMEROCALLIS 'BRIDGET'
RHODODENDRON 'BRIGITTE'
DELPHINIUM 'BRUCE'
HELENIUM 'BRUNO'

---

beesianus named for Bees, the
  Cheshire nursery and seed
  supplier
  Primula beesiana
belgicus from Belgium or the
  Netherlands
  Lonicera periclymenum 'Belgica'
  (early Dutch honeysuckle)
belladonna beautiful lady (referring
  to the one-time use of deadly
  nightshade as an eye cosmetic)
  Atropa bella-donna

Banksia named for Sir Joseph Banks (see page 34)
Baptisia from Greek *bapto*, to dye
Begonia named for Michel Bégon (1638–1710), French Canadian governor
Bellis from Latin *bellus*, pretty
Bergenia named for Karl August von Bergen (1704–1760), German professor
Bidens from Latin *bi-*, two, and *dens*, tooth
Bougainvillea named for Louis de Bougainville (1729–1811), French explorer
Brachyscome from Greek *brachys*, short, and *kome*, hair
Brunnera named for Samuel Brunner (1790–1844), Swiss botanist
Buddleja named for the Rev. Adam Buddle (1660–1715), British botanist and clergyman

bellidifolius with leaves like *Bellis* (daisy)
   Limonium bellidifolium
bellus beautiful
benedictus well spoken of; blessed
   Cnicus benedictus
bengalensis, benghalensis from Bengal
benzoin an aromatic gum
   Lindera benzoin
berolinensis from Berlin
   Populus × *berolinensis*

berthelotii named for a French consul and naturalist, Sabin Berthelot (1794–1880)
   Lotus berthelotii
bertolonii named for an Italian botanist, Antonio Bertoloni (1775–1869)
   Aquilegia bertolonii
betaceus like beet
betinus purple like beetroot
betonicifolius with leaves like *Stachys officinalis* (betony)
   Meconopsis betonicifolia
betulifolius with leaves like *Betula* (birch)
   Campanula betulifolia
bhutanicus from Bhutan
   Malus bhutanica

COMMON DESCRIPTIVE TERMS
bicolor bicoloured
   Dietes bicolor
   Eucomis bicolor
   Galega bicolor
   Kniphofia bicolor

bicornis with two horns
   Matthiola bicornis
bicuspidatus with two points
bidentatus with two teeth
biennis biennial
   Oenothera biennis
bifidus divided, split into two parts
   Escallonia bifida

Baby blue-eyes *Nemophila menziesii*
Baby's breath *Gypsophila paniculata*
Bachelor's buttons *Ranunculus aconitifolius*
Balloon flower *Platycodon*
Balsam *Impatiens*
Bamboo *Fargesia, Phyllostachys, Pleioblastus, Pseudosasa* etc.
Banana *Musa*
Barberry *Berberis*
Barberton daisy *Gerbera jamesonii*
Barley *Hordeum*
Barrenwort *Epimedium*
Bay, bay laurel *Laurus nobilis*
Bear's breeches *Acanthus*
Beauty berry *Callicarpa*
Beauty bush *Kolkwitzia amabilis*

BARLEY

Bee balm *Monarda didyma*
Beech *Fagus sylvatica*
Bellflower *Campanula*
Bells of Ireland *Moluccella laevis*
Bergamot *Monarda didyma*
Betony *Stachys officinalis*
Bindweed *Convolvulus*
Birch *Betula*
Bishop weed *Aegopodium podagraria*
Bistort *Persicaria bistorta*
Black gum *Nyssa sylvatica*
Black mulberry *Morus nigra*
Black-eyed Susan *Rudbeckia fulgida; Rudbeckia hirta; Thunbergia alata*

Blackthorn *Prunus spinosa*
Bladder senna *Colutea*
Blanket flower *Gaillardia*
Blazing star *Liatris*
Bleeding heart *Dicentra spectabilis*
Bloodroot *Sanguinaria canadensis*
Bloody cranesbill *Geranium sanguineum*
Blue oat grass *Helictotrichon sempervirens*
Bluebell, English *Hyacinthoides non-scripta*
Bluebell (Australia) *Sollya heterophylla*; (Scotland) *Campanula rotundifolia;* (USA) *Mertensia, Penstemon*
Bluebell, Spanish *Hyacinthoides hispanica*
Blueberry *Vaccinium*
Blueblossom *Ceanothus thyrsiflorus*
Blue-eyed grass *Sisyrinchium graminoides*
Blue-eyed Mary *Omphalodes verna*
Bog myrtle *Myrica gale*
Bogbean *Menyanthes trifoliata*
Boston ivy *Parthenocissus tricuspidata*
Bottlebrush *Callistemon*
Box elder *Acer negundo*
Box, Boxwood *Buxus*
Bramble *Rubus fruticosus*
Bridal wreath *Spiraea* 'Arguta'
Broom *Cytisus; Genista; Spartium*
Buckeye *Aesculus*
Buckthorn *Rhamnus*
Bugle *Ajuga reptans*
Bulrush *Typha*
Busy lizzie *Impatiens*
Butcher's broom *Ruscus aculeatus*
Buttercup *Ranunculus*
Butterfly bush *Buddleja davidii*

CULTIVAR NAMES
Strange but True
B

Iris 'Baboon Bottom'

Pelargonium 'Baby Bird's Egg'

Diascia 'Baby Bums'

Pelargonium 'Baby Snooks'

Chamaecyparis obtusa 'Bambi'

Verbascum 'Banana Custard'

Iris 'Bang'

Luzula sylvatica 'Barcode'

Rhododendron 'Bashful'

Aubrieta 'Belisha Beacon'

Hemerocallis 'Bette Davis Eyes'

Stachys byzantina 'Big Ears'

Fargesia murielae 'Bimbo'

Lupinus 'Bishop's Tipple'

Dianthus Black & White Minstrels Group

Iris 'Black Tie Affair'

Malus domestica (apple) 'Bloody Butcher'

Campanula persicifolia 'Blue Bloomers'

Larix kaempferi 'Blue Rabbit Weeping'

Pelargonium 'The Boar'

Geranium 'Bob's Blunder'

Iris 'Boo'

Houttuynia cordata 'Boo-Boo'

Fuchsia 'Bow Bells'

Ranunculus ficaria 'Brazen Hussy'

Phlox paniculata 'Brigadier'

Pelargonium 'Brilliantine'

Lupinus 'Bubblegum'

Common descriptive terms

biflorus with two flowers
Abelia biflora
Camassia biflora
Crocus biflorus
Fritillaria biflora

---

bifolius with two leaves
Scilla bifolia
bifurcatus bifurcate; forked
bignonioides like *Bignonia*
Catalpa bignonioides
bilobus, bilobatus with two lobes
Ginkgo biloba
bipinnatus twice pinnate
Cosmos bipinnatus
bisectus divided into two equal parts
biserratus twice toothed (i.e. with serrated teeth)
bistortus twice twisted
Persicaria bistorta 'Superba'
biternatus twice ternate
bithynicus from Bithynia, a former region of Asia Minor
Fritillaria bithynica
blandus mild, not bitter; pleasing
Anemone blanda
blepharophyllus with leaves fringed like eyelashes
Arabis blepharophylla
bodinieri named for a French missionary and plant collector, Emile Bodinier (1842–1901)
Callicarpa bodinieri

bodnantensis named for Bodnant
Garden in North Wales
  Viburnum × *bodnantense*
bohemicus from the former kingdom
of Bohemia in Eastern Europe
  Geranium bohemicum
bolivianus, boliviensis from Bolivia
  Fuchsia boliviana
bombycifer with silky hairs
  Verbascum bombyciferum
bombycinus of silk
bonariensis from Buenos Aires
  Verbena bonariensis
bononiensis from Bologna in Italy
bonus-henricus 'good Henry'
  Chenopodium bonus-henricus (good
  King Henry)
borbonicus Latinized 'Bourbon':
  referring to the island of Réunion
  in the Indian Ocean (Ile Bourbon)
  or the Bourbon kings of France
borealis northern
borisii; borisii-regi named for King
  Boris of Bulgaria
  Abies borisii-regi
  Geum 'Borisii'
borneensis from Borneo
bracchiatus like arms; branched
  at right angles
brachy- short-
brachylobus with short lobes
  Artemisia brachyloba
brachysiphon with a short tube
  Hebe brachysiphon

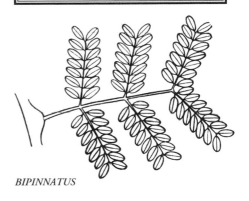

BIPINNATUS

brachystachyus with a short spike
  Kniphofia brachystachya
bracteatus with bracts
  Xerochrysum bracteatum
bracteosus with large, showy
  or significant bracts
brasiliensis from Brazil
brevifolius with short leaves
  Aloe brevifolia

CULTIVAR NAMES
Foreign Expressions
B

BARBAROSSA (*Italian*) RED BEARD
(*BERBERIS* × *CARMINEA* 'BARBAROSSA')

BARBE BLEU (*French*) BLUE BEARD
(*PELARGONIUM* 'BARBE BLEU')

BELLE ETOILE (*French*)
BEAUTIFUL STAR
(*PHILADELPHUS* 'BELLE ETOILE')

BENI-CHIDORI (*Japanese*)
RED THOUSAND BIRDS
(*PRUNUS MUME* 'BENI-CHIDORI')

BENI-SHIDARE (*Japanese*)
WEEPING RED
(*ACER PALMATUM* 'BENI-SHIDARE')

BERCEUSE (*French*) LULLABY
(*NARCISSUS* 'BERCEUSE')

BERGGOLD (*German*)
MOUNTAIN GOLD
(*ALYSSUM MONTANUM* 'BERGGOLD')

BLAUE DONAU (*German*)
BLUE DANUBE
(*CARYOPTERIS* × *CLANDONENSIS* 'BLAUE DONAU')

BLAUE GLOCKE (*German*)
BLUE BELL
(*PULSATILLA* 'BLAUE GLOCKE')

BLAUES MEER (*German*)
BLUE SEA
(*PULMONARIA ANGUSTIFOLIA* 'BLAUES MEER')

BLAUFUCHS (*German*) BLUE FOX
(*FESTUCA GLAUCA* 'BLAUFUCHS')

BLAUMEISE (*German*) BLUE TIT
(*HYDRANGEA MACROPHYLLA* 'BLAUMEISE')

BLAUSPIEGEL (*German*)
BLUE MIRROR
(*VERONICA PROSTRATA* 'BLAUSPIEGEL')

BLAUSTRUMPF (*German*)
BLUE STOCKING
(*MONARDA* 'BLAUSTRUMPF')

BLEKITNY ANIOL (*Polish*) BLUE ANGEL
(*CLEMATIS* 'BLEKITNY ANIOL')

BLEU NANTAIS (*French*) NANTES BLUE
(*CHAMAECYPARIS LAWSONIANA* 'BLEU NANTAIS')

BLÜTENTISCH (*German*)
BLOSSOM TABLE
(*HELENIUM* 'BLÜTENTISCH')

BOULE DE NEIGE (*French*)
SNOWBALL
(*CAMPANULA PERSICIFOLIA* 'BOULE DE NEIGE')

BRAUTSCHLEIER (*German*)
BRIDAL VEIL
(*ASTILBE* 'BRAUTSCHLEIER')

BRISE D'ANJOU (*French*)
ANJOU BREEZE
(*POLEMONIUM* BRISE D'ANJOU ('BLANJOU'))

BRONZESCHLEIER (*German*)
BRONZE VEIL
(*DESCHAMPSIA CESPITOSA* 'BRONZESCHLEIER')

brevipedunculatus with a short flower-stalk
   Ampelopsis brevipedunculata
brevipes with a short foot or stalk
   Malus brevipes
brevipetalus with short petals
   Hamamelis 'Brevipetala'
brevis short
brewerianus named for the American botanist William Brewer (1828–1910)
   Picea breweriana (Brewer's weeping spruce)
britannicus from Great Britain
bryoides like moss
bucarius, bucharicus from Bokhara in Turkestan
   Iris bucharica
buchananii named for John Buchanan (1819–1898), a Scottish botanist who worked in New Zealand
   Carex buchananii
bucinatus like a curved horn
bufonius relating to toads; growing in damp places
bulbifer bearing bulbs or bulbils
   Asplenium bulbiferum
bulbocodium with a woolly bulb
   Narcissus bulbocodium
bulbosus bulbous; swollen
bulgaricus from Bulgaria
bullatus blistered, bubbled, studded
   Cotoneaster bullatus

bulleyanus named for Arthur Bulley (1861–1942), a wealthy Liverpool merchant who sponsored plant hunters and founded Bees' nursery in Cheshire. Bulley's garden became the Liverpool Botanic Garden
   Primula bulleyana
bungeanus, bungei named for a Russian botanist, Alexander von Bunge (1803–1890)
   Euonymus bungeanus

COMMON DESCRIPTIVE TERMS

burkwoodii named for brothers Arthur (1888–1951) and Albert Burkwood, who ran Park Farm Nursery at Kingston upon Thames
   Daphne × *burkwoodii*
   Osmanthus × *burkwoodii*
   Viburnum × *burkwoodii*

---

burmanicus from Burma
bursa-pastoris shepherd's purse
   Capsella bursa-pastoris
buxifolius box-leaved
   Berberis buxifolia

COMMON DESCRIPTIVE TERMS

byzantinus from Istanbul, Turkey (formerly Byzantium)
   Colchicum byzantinum
   Galanthus plicatus subsp. *byzantinus*
   Gladiolus communis subsp. *byzantinus*
   Stachys byzantina

C

cachemirianus, cachemiricus from Kashmir (see also *cashmerianus*)
    Aralia cachemirica
    Lavatera cachemiriana
cadens falling
cadmicus metallic

COMMON DESCRIPTIVE TERMS
caeruleus deep sky blue
Allium caeruleum
Catananche caerulea
Molinia caerulea
Passiflora caerulea
Polemonium caeruleum

---

caesius light greyish blue
caespitosus, cespitosus clump-forming
    Deschampsia cespitosa
    Salvia caespitosa
caffer from South Africa
    Geranium caffrum
cairicus from Cairo
calabricus from Calabria in southern Italy
    Putoria calabrica
calamifolius with reedlike leaves
calamus reedlike
    Acorus calamus
calathinus cup-shaped; like a basket

calcaratus spurred
    Origanum calcaratum
calcareus of limestone; chalky white
    Polygala calcarea
calceiformis, calceolatus shaped like a little shoe or slipper
calcicola growing on limy soil
    Narcissus calcicola
calendulinus orange like *Calendula* (pot marigold)
calidus warm, hot

COMMON DESCRIPTIVE TERMS
californicus from California, USA
Aesculus californica
Brodiaea californica
Carpenteria californica
Erythronium californicum
Eschscholzia californica
Fremontodendron californicum
Sisyrinchium californicum
Zauschneria californica

---

calli-, calo- beautiful (see below)
callianthus with beautiful flowers
    Gladiolus callianthus
callicarpus with beautiful fruit
callimorphus beautifully shaped
callosus with a hard skin
    Saxifraga callosa
calthifolius with leaves like *Caltha* (marsh marigold)
    Geum calthifolium
calvus bald, hairless

CULTIVAR NAMES
Gardens Past and Present

PITTOSPORUM TENUIFOLIUM
'ABBOTSBURY GOLD'

POTENTILLA FRUTICOSA 'ABBOTSWOOD'

ORIGANUM VULGARE 'ACORN BANK'

HEMEROCALLIS 'APPLE COURT RUBY'

VIOLA 'BARNSDALE GEM'

LAVATERA × CLEMENTII 'BARNSLEY'

CARPENTERIA CALIFORNICA 'BODNANT'

HOSTA 'BRESSINGHAM BLUE'

ASTRANTIA 'BUCKLAND'

MAGNOLIA 'CAERHAYS BELLE'

PRUNUS LAUROCERASUS
'CASTLEWELLAN'

HELIOTROPIUM ARBORESCENS
'CHATSWORTH'

OLEA EUROPAEA
'CHELSEA PHYSIC GARDEN'

SALVIA PATENS 'CHILCOMBE'

EUPHORBIA GRIFFITHII 'DIXTER'

RHODODENDRON 'EXBURY NAOMI'

ERICA × DARLEYENSIS 'FURZEY'

SOLANUM CRISPUM 'GLASNEVIN'

CLEMATIS 'GRAVETYE BEAUTY'

ANEMONE HUPEHENSIS
'HADSPEN ABUNDANCE'

CONVALLARIA MAJALIS
'HARDWICK HALL'

CLEMATIS HARLOW CARR ('EVIP0004')

LAMIUM MACULATUM 'HATFIELD'

PENSTEMON 'HERGEST CROFT'

LAVANDULA ANGUSTIFOLIA 'HIDCOTE'

HELIANTHEMUM 'HIGHDOWN APRICOT'

SKIMMIA JAPONICA
'HIGHGROVE REDBUD'

VERBASCUM 'HYDE HALL SUNRISE'

PRIMULA 'INVEREWE'

CISTUS × DANSEREAUI 'JENKYN PLACE'

ABUTILON × SUNTENSE 'JERMYNS'

SKIMMIA × CONFUSA 'KEW GREEN'

ROSA FILIPES 'KIFTSGATE'

CAREX ELATA 'KNIGHTSHAYES'

ARTEMISIA ABSINTHIUM
'LAMBROOK SILVER'

MAGNOLIA SPRENGERI VAR. *DIVA*
'LANHYDROCK'

BUDDLEJA 'LOCHINCH'

LAVANDULA ANGUSTIFOLIA
'MUNSTEAD'

PENSTEMON 'MYDDELTON GEM'

EUCRYPHIA × NYMANSENSIS
'NYMANSAY'

FUCHSIA MAGELLANICA VAR. *MOLINAE*
'SHARPITOR'

GERANIUM × MAGNIFICUM 'ROSEMOOR'

CHAENOMELES × SUPERBA
'ROWALLANE'

NYSSA SYLVATICA 'SHEFFIELD PARK'

PULMONARIA 'SISSINGHURST WHITE'

GERANIUM 'SPINNERS'

SALVIA × JAMENSIS 'TREBAH'

WATSONIA 'TRESCO DWARF PINK'

CEANOTHUS ARBOREUS
'TREWITHEN BLUE'

BETULA UTILIS 'WAKEHURST PLACE'

CISTUS × CRISPATUS 'WARLEY ROSE'

NARCISSUS 'WATERPERRY'

PINUS SYLVESTRIS 'WESTONBIRT'

HELIANTHEMUM 'WISLEY PRIMROSE'

LILIUM CANDIDUM

calycinus with a prominent or lasting calyx
   Halimium calycinum
calystegioides like *Calystegia* (e.g. bindweed)
cambessedesii named for a French botanist, Jacques Cambessedes (1799–1863)
   Paeonia cambessedesii
cambricus Welsh
   Meconopsis cambrica (Welsh poppy)
camellifolius with leaves like *Camellia*
   Ilex × altaclerensis 'Camellifolia'
campaniflorus with bell-shaped flowers
   Clematis campaniflora

COMMON DESCRIPTIVE TERMS
campanulatus bell-shaped
   Agapanthus campanulatus
   Enkianthus campanulatus
   Penstemon campanulatus
   Persicaria campanulatus
   Rhododendron campanulatum

campestris of plains, flat land or fields
   Acer campestre
camphoratus like camphor
   Nepeta camphorata
camporum of plains
campylo- curved, bent
camtschatcensis, camtschaticus from Kamchatka, Siberia
   Lysichiton camtschatcensis
   Filipendula camtschatica

COMMON DESCRIPTIVE TERMS
canadensis from Canada or the north-eastern USA
   Amelanchier canadensis
   Cercis canadensis
   Cornus canadensis
   Populus × canadensis
   Sanguinaria canadensis
   Sanguisorba canadensis
   Tsuga canadensis

canaliculatus with a channel, like a pipe
   Narcissus canaliculatus

canariensis from the Canary Islands

Genista canariensis

Hedera canariensis

Lavandula canariensis

Phoenix canariensis

Pinus canariensis

Salvia canariensis

canarinus canary yellow

candicans becoming white

   Galtonia candicans

candidissimus very white

   Arisaema candidissimum

candidulus rather white

   Berberis candidula

candidus pure white

   Lilium candidum (Madonna lily)

canescens becoming greyish white

   Populus × *canescens*

caninus relating to dogs (used to
denote something inferior or
coarse)

   Rosa canina

cannabinus like hemp

Althaea cannabina

Datisca cannabina

Eupatorium cannabinum

Hibiscus cannabinus

cantabricus from Cantabria in Spain

   Daboecia cantabrica

CULTIVAR NAMES
Strange but True
C

IRIS SIBIRICA 'CAESAR'S BROTHER'

SEMPERVIVUM 'CAFÉ'

CALLUNA VULGARIS
'CALIFORNIA MIDGE'

GERANIUM PHAEUM 'CALLIGRAPHER'

HEUCHERA 'CAN-CAN'

PRIMULA 'CAPTAIN BLOOD'

HOSTA 'CAPTAIN KIRK'

RANUNCULUS REPENS 'CAT'S EYES'

MALUS DOMESTICA (APPLE)
'CATSHEAD'

CIRCAEA LUTETIANA 'CAVEAT EMPTOR'

PHLOX DIVARICATA SUBSP. *LAPHAMII*
'CHATTAHOOCHEE'

SAXIFRAGA 'CHEAP CONFECTIONS'

IRIS 'CHICKEN LITTLE'

NARCISSUS 'CHIT CHAT'

MENTHA 'CHOCOLATE PEPPERMINT'

PAPAVER ORIENTALE 'CHOIR BOY'

IRIS 'CHUBBY CHEEKS'

PHLOX PANICULATA 'CINDERELLA'

PANICUM VIRGATUM 'CLOUD NINE'

SEDUM CAUTICOLA 'COCA-COLA'

SEMPERVIVUM 'COBWEB CAPERS'

FUCHSIA 'COME DANCING'

PRIMULA AURICULA 'CONSERVATIVE'

PHLOX DOUGLASII 'CRACKERJACK'

PITTOSPORUM 'CRINKLES'

FUCHSIA 'CRINKLEY BOTTOM'

PELARGONIUM 'CROCODILE'

JUNCUS DECIPIENS 'CURLY-WURLY'

CROCOSMIA × *CROCOSMIIFLORA*
'CUSTARD CREAM'

cantabrigiensis relating to
Cambridge, England
Rosa 'Cantabrigiensis'

CULTIVAR NAMES
Colours: Yellow

CAMELLIA × *WILLIAMSII*
'JURY'S YELLOW'

CORNUS SERICEA 'BUDD'S YELLOW'

CROCOSMIA MASONIORUM
'ROWALLANE YELLOW'

DAHLIA 'YELLOW HAMMER'

EREMURUS 'YELLOW GIANT'

ERICA CARNEA 'WESTWOOD YELLOW'

GENISTA PILOSA 'YELLOW SPREADER'

HEMEROCALLIS 'YELLOW LOLLIPOP'

HOSTA 'YELLOW RIVER'

IRIS SIBIRICA 'DREAMING YELLOW'

KNIPHOFIA 'SUNNINGDALE YELLOW'

MAGNOLIA × *BROOKLYNENSIS*
'YELLOW BIRD'

NARCISSUS 'YELLOW
CHEERFULNESS'

OENOTHERA FRUTICOSA
'YELLOW RIVER'

PENSTEMON PINIFOLIUS
'MERSEA YELLOW'

PHILADELPHUS 'YELLOW CAB'

PHORMIUM 'YELLOW WAVE'

PHYGELIUS AEQUALIS
'YELLOW TRUMPET'

PITTOSPORUM TENUIFOLIUM
'MELLOW YELLOW'

POTENTILLA FRUTICOSA
'YELLOW BIRD'

PRIMULA 'LISMORE YELLOW'

RHODODENDRON 'YELLOW HAMMER'

---

canterburiensis from the Canterbury
Plains of New Zealand
Hebe canterburiensis

cantoniensis from Guangzhou
(Canton), China
Disporum cantoniense

capensis from the Cape (of Good
Hope), meaning South Africa
Phygelius capensis

caperatus wrinkled

capillaris hairlike, slender
Artemisia capillaris

capillifolius with hairlike leaves
Eupatorium capillifolium

capillipes slender-footed
Acer capillipes

capillus-veneris Venus' hair
Adiantum capillus-veneris

COMMON DESCRIPTIVE TERMS
capitatus with a solid head or tip
Cornus capitata
Persicaria capitata
Picea abies 'Capitata'
Primula capitata

---

cappadocicus from Cappadocia,
Turkey
Omphalodes cappadocica

caprea a nanny goat; relating to goats
Salix caprea (goat willow)

capreolatus with tendrils or
supports
Bignonia capreolata

Angelica *Angelica archangelica*
Apple mint *Mentha suaveolens*
Basil *Ocimum basilicum*
Bay *Laurus nobilis*
Borage *Borago officinalis*
Camomile *Chamaemelum nobile*
Caraway thyme *Thymus herba-barona*
Chervil *Anthriscus cerefolium*
Chinese chives *Allium tuberosum*
Chives *Allium schoenoprasum*
Coriander *Coriandrum sativum*
Corsican mint *Mentha requienii*
Costmary (Alecost) *Leucanthemum balsamita*
Dill *Anethum graveolens*

Eau de Cologne mint *Mentha × piperita* f. *citrata*
Fennel *Foeniculum vulgare*
Feverfew *Tanacetum parthenium*
Ginger mint *Mentha × gracilis*
Hyssop *Hyssopus officinalis*
Lavender *Lavandula*
Lemon balm *Melissa officinalis*
Lemon grass *Cymbopogon citratus*
Lemon thyme *Thymus × citriodorus*
Lemon verbena *Aloysia triphylla*
Lovage *Levisticum officinale*
Marjoram *Origanum* (see also Pot marjoram etc.)
Mint *Mentha* (see also Peppermint, Spearmint etc.)
Oregano *Origanum vulgare*
Parsley *Petroselinum crispum*
Pennyroyal *Mentha pulegium*
Peppermint *Mentha × piperita*
Pineapple mint *Mentha suaveolens*
Pot marjoram *Origanum onites*
Rosemary *Rosmarinus officinalis*
Rue *Ruta graveolens*
Sage *Salvia officinalis*
Salad burnet *Sanguisorba minor*
Sorrel *Rumex acetosa*
Southernwood *Artemisia abrotanum*
Spearmint *Mentha spicata*
Summer savory *Satureja hortensis*
Sweet cicely *Myrrhis odorata*
Sweet marjoram *Origanum majorana*
Tansy *Tanacetum vulgare*
Tarragon *Artemisia dracunculus*
Thyme *Thymus vulgaris* (see also Lemon thyme etc.)
Wild marjoram *Origanum vulgare*
Winter savory *Satureja montana*

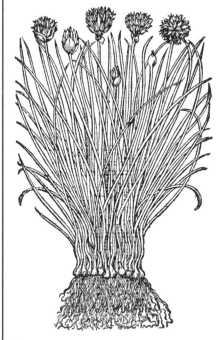

CHIVES

CULTIVAR NAMES
Personal Names
C

GERANIUM CINEREUM 'CAROL'

PRIMULA AURICULA 'CAROLE'

WISTERIA 'CAROLINE'

IRIS 'BROADLEIGH CAROLYN'

HEMEROCALLIS 'CATHY'S SUNSET'

AJUGA REPTANS 'CATLIN'S GIANT'

MALUS DOMESTICA (APPLE) 'CHARLOTTE'

PELARGONIUM 'CHERIE'

PELARGONIUM 'CHRISSIE'

RHODODENDRON 'CHRISTINA'

SAXIFRAGA 'CHRISTINE'

DIANTHUS 'CHRISTOPHER'

ERICA CINEREA 'CINDY'

VERBASCUM 'CLAIRE'

PRIMULA AURICULA 'CLARE'

PARAHEBE LYALLII 'CLARENCE'

PAEONIA 'CLAUDIA'

IRIS 'CLEO'

FUCHSIA 'CLIFF'S HARDY'

PENSTEMON 'CONNIE'S PINK

CROCOSMIA × CROCOSMIIFLORA 'CONSTANCE'

VIOLA 'CORDELIA'

RHODODENDRON 'CYNTHIA'

caprifolius with goatlike leaves

Lonicera caprifolium

capsicinus red like *Capsicum* (pepper)

capsularis like a capsule

capucinus red like *Tropaeolum* (nasturtium)

carbonaceus black like charcoal

cardia-, cardio- heart-

cardiacus relating to the heart

Leonurus cardiaca

cardinalis cardinal red

Lobelia cardinalis

cardiophyllus with heart-shaped leaves

Codonopsis cardiophylla

cardunculus like a little thistle

Cynara cardunculus

carens absent or lacking

caricus from Caria, in Asia Minor

Ficus carica

carinatus with a keel or shell

Allium carinatum

cariosus decayed

carlesii named for the British plant-collector William Carles (1867–1900)

Viburnum carlesii

carmesinus crimson

carmineus carmine

Metrosideros carminea

COMMON DESCRIPTIVE TERMS

carneus flesh-coloured

Erica carnea

Gladiolus carneus

Lychnis chalcedonica 'Carnea'

Polemonium carneum

---

carniolicus from Carniola, an area of the former Yugoslavia

Primula carniolica

carnosulus slightly fleshy
  Hebe carnosula
carnosus fleshy, succulent
  Hoya carnosa
carolinianus from North or South
  Carolina, USA
  Anemone caroliniana
carophyllus with fleshy leaves
  Dianthus carophyllus
carpaticus from the Carpathian
  Mountains
  Campanula carpatica
carpicus relating to fruits
carpinifolius with leaves like *Carpinus*
  (hornbeam)
  Acer carpinifolium
-carpus -fruit, -fruited
  Physocarpus
carthusianus, carthusianorum relating
  to Carthusian monks
  Dianthus carthusianorum
  Dryopteris carthusiana
carvifolius with leaves like *Carum*
  carvi (caraway)
  Erodium carvifolium
caryo- nut

COMMON DESCRIPTIVE TERMS
cashmerianus from Kashmir
Campanula cashmeriana
Corydalis cashmeriana
Cupressus cashmeriana
Phlomis cashmeriana
Polemonium cashmerianum

CULTIVAR NAMES
British Towns and Cities

FUCHSIA 'COUNTESS OF ABERDEEN'
PITTOSPORUM 'ARUNDEL GREEN'
MALUS DOMESTICA (APPLE)
  'BEAUTY OF BATH'
FUCHSIA 'BRIGHTON BELLE'
PASSIFLORA 'STAR OF BRISTOL'
CLEMATIS 'BURFORD WHITE'
MONARDA 'CAMBRIDGE SCARLET'
DIANTHUS 'CONWY STAR'
ERICA × *VEITCHII* 'EXETER'
MAGNOLIA GRANDIFLORA 'EXMOUTH'
CLEMATIS 'ELEANOR OF GUILDFORD'
DIANTHUS
  'HEREFORD BUTTER MARKET'
GERANIUM SANGUINEUM 'INVERNESS'
DIANTHUS 'IPSWICH PINK'
SALIX CAPREA 'KILMARNOCK'
AGAPANTHUS 'KINGSTON BLUE'
FUCHSIA 'CITY OF LEICESTER'
CLEMATIS 'LINCOLN STAR'
DAHLIA 'BISHOP OF LLANDAFF'
RANUNCULUS FICARIA
  'NEWTON ABBOT'
AGAPANTHUS CAMPANULATUS
  'OXFORD BLUE'
CALLUNA VULGARIS 'RADNOR'
ROSA 'SOUTHAMPTON'
HEMEROCALLIS 'STAFFORD'
CYTISUS × *PRAECOX* 'WARMINSTER'
LONICERA PERICLYMENUM
  'WINCHESTER'
CLEMATIS 'BELLE OF WOKING'
CARYOPTERIS × *CLANDONENSIS*
  'WORCESTER GOLD'

cassideus shaped like a helmet

cassus empty, hollow, devoid of

castaneifolius with leaves like
  Castanea (chestnut)
  Quercus castaneifolia

castaneus chestnut-coloured

castellanus relating to castles

catalpifolius with leaves like
  Catalpa
  Acer longipes subsp. *catalpifolium*

catarius of cats
  Nepeta cataria

catarractae of waterfalls
  Parahebe catarractae

catenatus like a chain

catharticus purgative, purifying
  Rhamnus cathartica

cathayanus from China
  Cardiocrinum cathayanum

catilliformis shaped like a saucer

---

caucasicus from the Caucasus

Arabis alpina subsp. *caucasica*

Artemisia caucasica

Corydalis caucasica

Fritillaria caucasica

Veronica caucasica

---

caudatifolius with tail-like leaves
  Acer caudatifolium

caudatus with a tail
  Amaranthus caudatus

caudiformis shaped like a tail

---

caulescens with a stem

Agapanthus caulescens

Asarum caulescens

Kniphofia caulescens

Mandragora caulescens

---

caulinus of or on the stem

-caulis -stemmed

cavernosus full of cavities or hollows

cavus hollow
  Corydalis cava

celatus concealed

cellulosus cellular

celsus high

cembra Italian name for Arolla pine
  Pinus cembra

centi-, centum hundred- (sometimes used to mean 'too many to count easily', as in centipede)
  Rosa × *centifolia*

cephallenicus, cephalonicus from the Greek island of Cephallonia
  Abies cephalonica
  Campanula garganica subsp. *cephallenica*

-cephalus head

ceraceus waxy

cerasifer cherry-bearing
  Prunus cerasifera (cherry plum)

cerasinus cherry red

cerastioides Like *Cerastium* (snow-in-summer; mouse-ear)
  Gypsophila cerastioides

Cabbage palm *Cordyline*
Calico bush *Kalmia*
California buckeye *Aesculus californica*
Californian lilac *Ceanothus*
Californian poppy *Eschscholzia*
Californian tree poppy *Romneya*
Camomile *Chamaemelum nobile*
Canary creeper *Tropaeolum peregrinum*
Candytuft *Iberis*
Canterbury bells *Campanula medium*
Cape figwort *Phygelius capensis*
Caper spurge *Euphorbia lathyris*
Cardinal flower *Lobelia cardinalis*
Cardoon *Cynara cardunculus*
Carnation *Dianthus*
Castor oil plant *Ricinus*
Catmint, catnip *Nepeta*
Cedar of Lebanon *Cedrus libani*
Celandine, greater *Chelidonium majus*
Celandine, lesser *Ranunculus ficaria*
Chatham Island forget-me-not
 Myosotidium hortensia
Cheddar pink *Dianthus*
 gratianopolitanus
Cherry *Prunus*

Cherry laurel *Prunus laurocerasus*
Cherry plum *Prunus cerasifera*
Cherry-pie *Heliotropium arborescens*
Chestnut, horse *Aesculus hippocastanum*
Chestnut, sweet *Castanea sativa*
Chilean glory flower *Eccremocarpus*
 scaber
Chimney bellflower *Campanula*
 pyramidalis
China aster *Callistephus*
Chinese lanterns *Physalis alkekengi*
Chocolate vine *Akebia quinata*
Chokeberry *Aronia*
Christmas box *Sarcococca*
Christmas rose *Helleborus niger*
Chusan palm *Trachycarpus fortunei*
Cider gum *Eucalyptus gunnii*
Cinquefoil *Potentilla*
Clary *Salvia viridis* var. *comata*
Cockspur thorn *Crataegus crus-galli*
Coleus *Solenostemon*
Columbine *Aquilegia*
Comfrey *Symphytum*
Coneflower *Echinacea*; *Rudbeckia*
Coral bells, Coral flower *Heuchera*
Coral-bark maple *Acer palmatum*
 'Sango-kaku'
Corkscrew hazel *Corylus avellana*
 'Contorta'
Cornelian cherry *Cornus mas*
Cornflower *Centaurea cyanus*
Cowslip *Primula veris*
Cranesbill *Geranium*
Crown imperial *Fritillaria imperialis*
Cuckoo flower *Cardamine pratensis*
Curry plant *Helichrysum italicum*
Cypress *Cupressus*
Cypress spurge *Euphorbia cyparissias*

LESSER CELANDINE

ceratocarpus with horned fruits
  Euphorbia ceratocarpa
cercidifolius with leaves like *Cercis*
  Disanthus cercidifolius
cerealis related to farming
cerefolius with waxy leaves
  Anthriscus cerefolium (chervil)

COMMON DESCRIPTIVE TERMS

cernuus drooping, downturned
  Allium cernuum
  Lilium cernuum
  Primula cernua
  Pulsatilla cernua

cervinus fawn-coloured
  Mentha cervina
cespitosus (see *caespitosus*)
cevennensis from the Cevennes
  in southern France
  Pulmonaria longifolia subsp.
  cevennensis
chaetocarpus with very hairy fruits
  Lonicera chaetocarpa
chaixii named for Dominique Chaix
  (1730–1799), a French botanist
  Verbascum chaixii
chalcedonicus from Chalcedon,
  a former region of Asia Minor
  Lychnis chalcedonica
chamae- low-growing, on the ground
chamaecyparissus like *Chamaecyparis*,
  false cypress
  Santolina chamaecyparissus

chamaemelifolius with leaves like
　Chamaemelum (camomile)
　Artemisia chamaemelifolia
chasmanthus with wide-open flowers
chathamicus from the Chatham
　Islands in the South Pacific
　Astelia chathamica
cheilanthus with lipped flowers
chilensis from Chile
　Puya chilensis
chiloensis from Chiloé, an island
　off the west coast of Chile
　Fragaria chiloensis
-chilus -lipped

CILIATUS

COMMON DESCRIPTIVE TERMS

chinensis Chinese (see also *sinensis*)
Abelia chinensis
Astilbe chinensis
Betula chinensis
Cornus kousa var. *chinensis*
Corylus chinensis
Enkianthus chinensis
Juniperus chinensis
Schisandra chinensis

chionanthus with snow-white flowers
　Primula chionantha
chionophilus snow-loving
　Ornithogalum chionophilum
-chiton covering
　Rhodochiton
chloranthus with green flowers
　Lathyrus chloranthus

chlorinus yellow-green
chloro- green
chloropetalus with green petals
　Trillium chloropetalum
chrom-, -chromus, -chrous relating
　to colour; coloured
　Euphorbia polychroma
chrys- gold
chrysanthus with golden flowers
　Crocus chrysanthus
chrysocomus with golden hairs
　Achillea chrysocoma
chrysographes marked with gold
　Iris chrysographes
chrysolepis with golden scales
chrysophyllus with golden leaves
　Phlomis chrysophylla

COMMON DESCRIPTIVE TERMS

ciliaris, ciliatus, ciliosus fringed
　with hairs (like an eyelid)
Bergenia ciliata
Chrysocoma ciliata
Erica ciliaris
Lysimachia ciliata
Melica ciliata
Mertensia ciliata

Callicarpa from Greek *kalli-*, beautiful, and *karpos*, fruit

Callistephus from Greek *kalli-*, beautiful, and *stephos*, crown

Calluna from Greek *kalluno*, to clean (heather was used to make brooms)

Camellia named for Georg Kamel (1661–1706), a pharmacist and botanical author working in the Philippines

Campanula from Latin *campana*, bell

Caryopteris from Greek *karyon*, nut, and *pteron*, wing

Chimonanthus from Greek *cheimon*, winter, and *anthos*, flower

Chionodoxa from Greek *chion*, snow, and *doxa*, glory

Chrysanthemum from Greek *chrysos*, gold, and *anthos*, flower

Claytonia named for John Clayton (1686–1773), an American botanist

Colchicum from Colchis, a region of the Caucasus

Convolvulus from Latin *convolvo*, to twine round

Coreopsis from Greek *koris*, a bug or tick, and *opsis*, resemblance

Coronilla from Latin *corona*, crown

Corylopsis from Greek *korylos*, hazel, and *opsis*, resemblance

Cosmos in Greek, beautiful

Crocus from Greek *krokos*, saffron

Cryptomeria from Greek *krypto*, to conceal, and *meris*, a part

Cyclamen from Greek *kyklos*, a circle

Cynoglossum from Greek *kyon*, dog, and *glossa*, tongue

cilicicus from Cilicica, a region of Asia Minor
  Cyclamen cilicicum

cineraria ash-coloured
  Senecio cineraria

cinereus like ashes, ash-grey
  Geranium cinereum

cinnamomeus cinnamon-brown
  Osmunda cinnamomea

circinatus coiled, circular
  Acer circinatum

cirratus, cirrhosus with tendrils
  Arthropodium cirratum
  Clematis cirrhosa

cirrhifolius with tendril-like leaves
  Polygonatum cirrhifolium

cissifolius with leaves like *Cissus*, a vine-like climber
  Acer cissifolium

citratus like *Citrus*
  Cymbopogon citratus

COMMON DESCRIPTIVE TERMS

citrinus lemon-yellow
  Callistemon citrinus
  Coronilla valentina subsp. *glauca* 'Citrina'
  Erythronium citrinum
  Hemerocallis citrina
  Iris foetidissima var. *citrina*
  Kniphofia citrina

---

citriodorus lemon-scented
  Thymus × *citriodorus*

citroides like *Citrus*

clad-, -cladus branch, shoot

clandestinus hidden

   Panicum clandestinum

clandonensis named for Clandon Park, Surrey

   Caryopteris × *clandonensis*

clarkei named for C.B. Clarke (1832–1906), superintendent of Calcutta Botanic Gardens

   Geranium clarkei

clathratus like a lattice or trellis

clausus closed

   Hosta clausa var. *normalis*

clavatus, claviformis club-shaped

clavellatus like a small club

clematideus like *Clematis*

   Codonopsis clematidea

clethroides like *Clethra* (sweet pepper bush)

   Lysimachia clethroides

clivorum of slopes

clusianus, clusii named for a Flemish botanist, Carolus Clusius (1526–1609)

   Gentiana clusii

   Tulipa clusiana

clypeolatus rather like a *clipeus,* a small circular shield used by the Romans

   Achillea clypeolata

coactus felted

coarctatus pressed closely together

cobaltinus cobalt blue

COMMON DESCRIPTIVE TERMS

coccineus deep red

Hedychium coccineum

Hibiscus coccineus

Pyracantha coccinea

Salvia coccinea

Schizostylis coccinea

CULTIVAR NAMES
Foreign Expressions
C, D

CALIENTE (*Spanish*) HOT
(*IRIS* 'CALIENTE')

CHITOSEYAMA (*Japanese*)
1,000-YEAR-OLD MOUNTAIN
(*ACER PALMATUM*
'CHITOSEYAMA')

CLAIR DE LUNE (*French*)
MOONLIGHT
(*PAEONIA* 'CLAIR DE LUNE')

DIAMANT (*German*)
DIAMOND
(*ASTILBE* 'DIAMANT')

DU MAÎTRE D'ECOLE (*French*)
OF THE SCHOOLMASTER
(*ROSA* 'DU MAÎTRE D'ECOLE')

DUNKELPRACHT (*German*)
DARK BEAUTY
(*HELENIUM* 'DUNKELPRACHT')

DUNKELSTE ALLER (*German*)
DARKEST OF ALL
(*ERIGERON* 'DUNKELSTE ALLER')

DÜSTERLOHE (*German*)
DARK FLAME
(*PHLOX PANICULATA* 'DÜSTERLOHE')

cochlearifolius with spoon-shaped
leaves
  Campanula cochlearifolia
cochleatus shaped like a snail-shell,
in a spiral
  Cotoneaster cochleatus
cockburnianus named for the
Cockburn family, living in China
  Rubus cockburnianus
-codon bell
  Platycodon
coelestinus sky blue
  Eupatorium coelestinum
coeruleus blue
  Puya coerulea
cognatus related
  Lychnis cognata
cognitus known, understood
-cola -dweller (e.g. *monticola,*
mountain-dweller)
colchicus from Colchis, a region
of Georgia
  Hedera colchica
collapsus collapsed
collinus of hills
  Geranium collinum
colombinus like a dove
coloratus coloured
  Pseudowintera colorata
colubrinus like a snake
columbianus from British Columbia
or the Columbia River, north-
western North America
  Lilium columbianum

columnaris columnar
  Acer platanoides 'Columnaris'
comans hairy; leafy
  Carex comans
comatus tufted
  Salvia viridis var. *comata*
commiscens intermingling
commixtus mixed together
  Sorbus commixta

COMMON DESCRIPTIVE TERMS

communis general, universal
Ferula communis
Gladiolus communis subsp. *byzantinus*
Juniperus communis
Myrtus communis
Pyrus communis
Ricinus communis

MYRTUS COMMUNIS

WHAT'S IN A NAME?

BOTANISTS AND NATURALISTS

banksii Sir Joseph Banks (1743–1820)

battandieri Jules Battandier (1848–1922)

brewerianus William Brewer (1828–1910)

buchananii John Buchanan (1819–1898)

bungeanus, bungei Alexander von Bunge (1803–1890)

clusianus, clusii Carolus Clusius (1526–1609)

coulteri Thomas Coulter (1793–1843)

dammeri Carl Dammer (1860–1920)

darwinii Charles Darwin (1809–1882)

delavayi, delavayanus Jean Marie Delavay (1834–1895)

fontanesianus, fontanesii René Desfontaines (1750–1833)

grayi Asa Gray (1810–1888)

gunnii Ronald Gunn (1808–1881)

hookeri Sir William Hooker (1785–1865) or son Sir Joseph (1817–1911)

jacquemontii, jacquemontianus Victor Jacquemont (1801–1832)

juddii William Judd (1888–1946)

lamarckii Chevalier Jean-Baptiste de Monet Lamarck (1744-1829)

langsdorffii Georg Langsdorf (1774–1852)

lavalleei Pierre Lavalle (1836–1884)

linnaeoides Carl Linnaeus (1707–1778)

ludlowii Frank Ludlow (1885–1972)

maackii Richard Maack (1825–1886)

makinoi Tomitaro Makino (1863–1957)

maximowiczianus, maximowiczii Karl Maximowicz (1827–1891)

menziesii Archibald Menzies (1754–1842)

mlokosewitschii Ludwik Mlokosiewicz (1831–1909)

nuttallii Thomas Nuttall (1786–1859)

perralderianus Henri de la Perraudière (1831–1861)

sargentii, sargentianus Charles Sprague Sargent (1841–1927)

schillingii Tony Schilling

selloanus Friedrich Sello or Sellow (1789–1831)

sibthorpianus, sibthorpii Humphrey Sibthorp (1713–1797) or his son John (1758–1796)

sieboldianus, sieboldii Philipp von Siebold (1796–1866)

spooneri Herman Spooner (1878–1976)

stewartii Laurence Stewart (1877–1934)

thunbergii Carl Thunberg (1743–1828)

tommasinianus Muzio de' Tommasini (1794–1879)

---

commutatus changing; close to another species

  Papaver commutatum (ladybird poppy)

comosus hairy, tufted

  Muscari comosum

COMMON DESCRIPTIVE TERMS

compactus compact, dense

  Abies concolor 'Compacta'

  Brachyglottis compacta

  Deutzia compacta

  Viburnum opulus 'Compactum'

complexus complex; encircled
complicatus complex
  Rosa 'Complicata'
compositus compound
  Erigeron compositus
compressus flattened
  Juniperus communis 'Compressa'
concatenatus linked together
  Cardamine concatenata
concholobus with lobes like sea-shells
  Primula concholoba
concinnatum, concinnus well
  arranged, neat
  Arum concinnatum
  Arisaema concinnum
concolor the same colour all over; of
  the same colour
  Abies concolor
condensatus condensed, closely
  packed
  Phlox condensata
conditus preserved, stored
confertus crowded
  Penstemon confertus
conflatus united, fused together
conformis of the same shape

COMMON DESCRIPTIVE TERMS

confusus uncertain; easily mistaken;
  mingled
Iris confusa
Mahonia confusa
Sarcococca confusa
Skimmia × *confusa*

---

congestiflorus with closely packed
  flowers
  Lysimachia congestiflora
congestus congested, closely packed
  together
  Cotoneaster congestus
conglomeratus conglomerate,
  clustered
  Hedera helix 'Conglomerata'
congolanus Congolese
conicus cone-shaped
  Carex conica
conifer cone-bearing
  Manglieta conifera
conjugatus joined in twos
conjunctus joined
conoideus like a cone
consanguineus related
  Arisaema consanguineum
consimilis similar in every way,
  completely alike
conspicuus conspicuous
  Cotoneaster conspicuus
constantinopolitanus from Istanbul
  (formerly Constantinople), Turkey
  Ranunculus constantinopolitanus
contaminatus contaminated,
  impure
  Lachenalia contaminata
contiguus adjoining
continentalis continental
  Aralia continentalis
contortus twisted, contorted
  Corylus avellana 'Contorta'

controversus turned against, lying
   opposite; disputed
   Cornus controversa
convexus convex
   Ilex crenata 'Convexa'
convolutus rolled up lengthways
   Sisyrinchium convolutum
convolvulaceus like *Convolvulus*
   (bindweed etc.)
   Codonopsis convolvulacea
cookianum named for the English
   navigator Captain James Cook
   (1728–1779)
   Phormium cookianum
corallinus coral-coloured
   Berberidopsis corallina
coralloides like coral
   Ozothamnus coralloides

COMMON DESCRIPTIVE TERMS
cordatus, cordiformis heart-shaped
Alnus cordata
Aralia cordata
Carya cordiformis
Houttuynia cordata
Macleaya cordata
Pontederia cordata

cordifolius with heart-shaped leaves
Aster cordifolius
Bergenia cordifolia
Crambe cordifolia
Eucryphia cordifolia
Globularia cordifolia

---

coreanus from Korea
   Carpinus coreana
coriaceus, coriarius leathery
   Holboellia coriacea
corifolius with leathery leaves
corniculatus with little horns
   Glaucium corniculatum
corniger horned
   Euphorbia cornigera
-cornis -horned
cornubiensis from Cornwall, England
   Polypodium interjectum
   'Cornubiense'

CULTIVAR NAMES
Fictional Characters
PRIMULA AURICULA 'BILBO BAGGINS'
CLEMATIS 'DANIEL DERONDA'
LIGULARIA DENTATA 'DESDEMONA'
DAHLIA 'LITTLE DORRIT'
MALUS DOMESTICA (APPLE) 'FALSTAFF'
PRIMULA AURICULA 'ELI JENKINS'
ROSA 'LONG JOHN SILVER'
CLEMATIS 'LITTLE NELL'
CLEMATIS 'NIOBE'
PIERIS JAPONICA 'SCARLETT O'HARA'
PELARGONIUM 'SANCHO PANZA'
PAPAVER ORIENTALE 'PETER PAN'
× *PHYLLIOPSIS HILLIERI* 'PINOCCHIO'
HOSTA 'POOH BEAR'
PHLOX PANICULATA 'PROSPERO'
TULIPA 'RED RIDING HOOD'
NARCISSUS 'RIP VAN WINKLE'
PELARGONIUM 'JUST WILLIAM'

JUNIPERUS HORIZONTALIS 'ANDORRA COMPACT'

DAHLIA 'ARABIAN NIGHT'

CANNA 'AUSTRALIA'

NERINE 'KING OF THE BELGIANS'

RHODODENDRON 'BRAZIL'

GERANIUM MACRORRHIZUM 'BULGARIA'

CLEMATIS 'BURMA STAR'

HEMEROCALLIS 'CANADIAN BORDER PATROL'

MISCANTHUS SINENSIS 'CHINA'

FUCHSIA 'ROSE OF DENMARK'

BRUGMANSIA VERSICOLOR 'ECUADOR PINK'

PELARGONIUM 'LA FRANCE'

ASTILBE 'DEUTSCHLAND'

PHLOX PANICULATA 'MISS HOLLAND'

CLEMATIS 'STAR OF INDIA'

NARCISSUS 'EMPRESS OF IRELAND'

CANNA 'ITALIA'

ARGYRANTHEMUM 'JAMAICA PRIMROSE'

HEMEROCALLIS 'MOROCCO RED'

HEBE ODORA 'NEW ZEALAND GOLD'

CLEMATIS 'POLISH SPIRIT'

EUPHORBIA CHARACIAS 'PORTUGUESE VELVET'

PENSTEMON 'RUSSIAN RIVER'

HEMEROCALLIS 'SCOTLAND'

PELARGONIUM 'SPANISH ANGEL'

NARCISSUS 'TIBET'

FUCHSIA 'TURKISH DELIGHT'

PENSTEMON 'WELSH DAWN'

PRIMULA AURICULA 'ZAMBIA'

cornutus horned
Viola cornuta

COMMON DESCRIPTIVE TERMS
coronarius used for wreaths
  or garlands
Anemone coronaria
Hedychium coronarium
Lychnis coronaria
Malus coronaria
Philadelphus coronarius

---

corrugatus corrugated, furrowed

COMMON DESCRIPTIVE TERMS
corsicus from the island of Corsica
Acinos corsicus
Cistus creticus subsp. *corsicus*
Colchicum corsicum
Crocus corsicus
Erodium corsicum

---

corymbiflorus with flowers arranged
  in a corymb
  Deutzia setchuenensis var.
  corymbiflora
corymbosus corymbose; with corymbs
  Vaccinium corymbosum
costatus ribbed
  Betula costata
cosyrensis from Pantellaria (formerly
  Cosyra), a small island between
  Sicily and North Africa
  Limonium cosyrense

coulteri named for Thomas Coulter (1793–1843), an Irish botanist and plant collector
 Romneya coulteri
coum from the Greek island of Kos
 Cyclamen coum

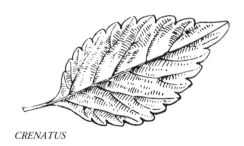

COMMON DESCRIPTIVE TERMS
crassifolius with thick leaves
Bergenia crassifolia
Fritillaria crassifolia
Gladiolus crassifolius
Ilex aquifolium 'Crassifolia'
Pittosporum crassifolium
Pseudopanax crassifolius

crassus thick
creber pressed closely together
crenatus with rounded teeth
 Ilex crenata
crenulatus with small rounded teeth
 Eucalyptus crenulata
crescens growing
cretaceus chalky white

COMMON DESCRIPTIVE TERMS
creticus from Crete
Campanula cretica
Cistus creticus
Cyclamen creticum
Eryngium creticum

crinitus covered with long hairs
 Carex crinita

crispulus curly-haired
 Hosta crispula

COMMON DESCRIPTIVE TERMS
crispus, crispatus curled
Asplenium scolopendrium Crispum Group
Buddleja crispa
Cistus × *crispatus* 'Warley Rose'
Ilex aquifolium 'Crispa'
Mentha spicata var. *crispa*
Petroselinum crispum (parsley)
Solanum crispum
Teucrium scorodonia 'Crispum Marginatum'

crista-galli a cockerel's comb
 Erythrina crista-galli
cristatellus with a small crest
 Carex cristatella
cristatus crested
 Iris cristata
crithmoides like *Crithmum* (rock samphire)
 Inula crithmoides

WHAT'S IN A NAME?

GARDENS AND GARDENERS

bodnantensis Bodnant Garden in North Wales

clandonensis Clandon Park, Surrey

elwesii Henry John Elwes (1846–1922), British plantsman

ericsmithii Eric Smith (1917–1986), plantsman and gardener at Hadspen, Somerset

fletcheri Harold Fletcher (1907–1978), Director of RHS Garden Wisley and Keeper of the Royal Botanic Garden Edinburgh

kewensis the Royal Botanic Gardens at Kew, London

lyonii John Lyon (1765–1814), Scottish gardener and botanist

nymansensis Nymans Garden, West Sussex

poscharskyanus Gustav Poscharsky (1832–1914), German gardener

robinsonianus William Robinson (1839–1935), Irish-born gardener, designer and influential plantsman (see page 173)

sternianus, sternii Sir Frederick Stern (1884–1967), plantsman, author and creator of a remarkable chalk garden at Highdown near Worthing, Sussex

tradescantii John Tradescant (1570–1638) and his son (1608–1662) of the same name, English travellers, plantsmen and royal gardeners

warleyensis, willmottianum Ellen Willmott (1858–1934), plantswoman and creator of an extravagant garden at Warley Place, Essex

croaticus from Croatia
  Helleborus croaticus

crocatus, croceus saffron yellow
  Iris crocea
  Tritonia crocata

cruciatus, cruciformis cross-shaped
  Gentiana cruciata

cruentus bloody
  Dianthus cruentus

crus-galli a cockerel's spur
  Crataegus crus-galli

cucullarius like a hood
  Dicentra cucullaria

cucullatus hooded
  Pelargonium cucullatum

cultorum of cultivated land
  Aubrieta × *cultorum*

cuneatus, cuneiformis cuneate (wedge-shaped)
  Prostanthera cuneata

cuneifolius with wedge-shaped leaves
  Saxifraga cuneifolia

cupanianus named for an Italian monk, Francesco Cupani (1657–1711)
  Anthemis punctata subsp. *cupaniana*

cupressoides like *Cupressus* (cypress)
  Hebe cupressoides

cupreus coppery
  Helianthemum cupreum

curassavicus from Curaçao in the Caribbean
  Asclepias curassavica

curviflorus with curved flowers

curvistylus with a curved style
  Polygonatum curvistylum
curvulus slightly curved
  Eragrostis curvula
cuspidatus cuspidate; with a sharp
  point
  Taxus cuspidata
cyaneus, cyanus dark blue
  Centaurea cyanus (cornflower)
cyathiformis cup-shaped
cyclamineus like *Cyclamen*
  Narcissus cyclamineus
cycloglossus with tongues arranged
  in a circle
  Iris cycloglossa
cyclophyllus with leaves in a circle
  Helleborus cyclophyllus

COMMON DESCRIPTIVE TERMS
cylindricus, cylindraceus cylindrical
Anemone cylindrica
Heuchera cylindrica 'Greenfinch'
Imperata cylindrica 'Rubra'
Magnolia cylindrica
Plectranthus cylindraceus

---

cymosus with cymes
cyparissias like cypress
  Euphorbia cyparissias
cyprius from Cyprus
  Cistus × *cyprius*
cyrenaicus from Cyrenaica, a region
  of Libya
  Arum cyrenaicum

CULTIVAR NAMES
Islands of the World

HEMEROCALLIS 'BALI HAI'
FUCHSIA 'BERMUDA'
FUCHSIA 'BORA BORA'
NERIUM OLEANDER 'ISLE OF CAPRI'
HEMEROCALLIS
  'CHRISTMAS ISLAND'
RHODODENDRON 'CRETE'
FUCHSIA 'FALKLANDS'
RHODODENDRON CALOSTROTUM
  'GIGHA'
CANNA 'GRAN CANARIA'
CLEMATIS 'GUERNSEY CREAM'
IRIS ENSATA 'HOKKAIDO'
HOSTA 'IONA'
MAGNOLIA 'JERSEY BELLE'
ROSMARINUS OFFICINALIS
  'MAJORCA PINK'
FUCHSIA 'ISLE OF MULL'
FUCHSIA 'LINDISFARNE'
ROSA 'MANX QUEEN'
GERANIUM 'ORKNEY PINK'
MUSA BASJOO 'SAKHALIN'
MALUS DOMESTICA (APPLE)
  'SCILLY PEARL'
IRIS 'STATEN ISLAND'
NARCISSUS 'TAHITI'
PLUMBAGO AURICULATA
  'TOBAGO BLUE'
FUCHSIA 'TRESCO'
FRITILLARIA AFFINIS
  'VANCOUVER ISLAND'
GERANIUM × *CANTABRIGIENSE*
  'WESTRAY'
CEANOTHUS 'PERSHORE ZANZIBAR'

D

dactyl- finger- (see below)
dactylifer finger-bearing
  Phoenix dactylifera

COMMON DESCRIPTIVE TERMS

dahuricus, dauricus from Dauria
  (Siberia/Mongolia)
  Actaea dahurica
  Betula dahurica
  Gentiana dahurica
  Lilium dauricum

---

dalmaticus from Dalmatia
  Geranium dalmaticum
damascenus from Damascus in Syria
  Nigella damascena
dammeri named for a German
  botanist, Carl Dammer (1860–
  1920)
  Cotoneaster dammeri
daphnoides like Daphne
  Salix daphnoides
darwinii named after Charles
  Darwin (1809–1882), the British
  naturalist, traveller and author
  of Origin of Species
  Berberis darwinii

WHAT'S IN A NAME?
SEASONS, WEATHER AND TIMING

aestivalis, aestivus of summer
aizoon ever-living
algidus cold
annuus annual
aurorius of dawn, daybreak, morning
autumnalis autumnal, of autumn
biennis biennial
calidus warm, hot
chion- snow
chionophilus snow-loving
diurnus diurnal, flowering by day
ephemerus ephemeral, short-lived
flos-cuculi flowering when the cuckoo
  sings
fugax fleeting, transitory, ephemeral
glacialis from glaciers or cold places
hyemalis of winter
longaevus long-lived
majalis of May, flowering in May

matronalis named for Matronalia,
  a Roman festival on 1 March
meridionalis (flowering at) midday,
  noon
nivalis, niveus snowy, snow-like
noctiflorus night-flowering
nyctagineus night-flowering
perennis perennial
praecox precocious, developing early
semperflorens always flowering
sempervirens evergreen
senescens growing old
serotinus late
solaris sun-loving
tardiflorus late-flowering
tardivus, tardus late
transitorius passing, transitory
trimestris of three months
veris, vernalis, vernus vernal, of spring

dasy- shaggy, hairy
dascyladus with hairy branches
dasyphyllus with hairy leaves
  Sedum dasyphyllum
dauricus see *dahuricus*

COMMON DESCRIPTIVE TERMS

davidii, davidianum named for
  the French missionary and plant
  collector Abbé Jean Pierre Armand
  David (1826–1900)
Acer davidii
Buddleja davidii
Chrysosplenium davidianum
Epimedium davidii
Photinia davidiana
Sophora davidii
Viburnum davidii

---

dealbatus whitened, powdery white
  Acacia dealbata
deca- ten- (see below)
decapetalus with ten petals
  Helianthus decapetalus
decaphyllus with ten leaves or
  leaflets
deciduus deciduous
  Larix decidua
decipiens deceiving (e.g. where one
  plant is easy to mistake for
  another)
  Corydalis decipiens
declivis sloping downward
decolor discoloured, faded

CULTIVAR NAMES
Personal Names
D

IMPATIENS WALLERIANA 'DAPPER DAN'
FUCHSIA 'DANNY BOY'
RHODODENDRON 'DAVID'
PELARGONIUM 'DAVINA'
VIBURNUM × *BODNANTENSE* 'DAWN'
CAMELLIA × *WILLIAMSII* 'DEBBIE'
ACER PLATANOIDES 'DEBORAH'
VIBURNUM CARLESII 'DIANA'
HAMAMELIS × *INTERMEDIA* 'DIANE'
PRIMULA × *FORSTERI* 'DIANNE'
GERANIUM 'DILYS'
HEMEROCALLIS 'DOMINIC'
PAEONIA LACTIFLORA 'DOREEN'
DIANTHUS 'DORIS'
CROCUS CHRYSANTHUS 'DOROTHY'
RHODOHYPOXIS 'DOUGLAS'
NARCISSUS 'UNCLE DUNCAN'

COMMON DESCRIPTIVE TERMS

decoratus, decorus decorative
Hosta decorata
Iris decora
Osmanthus decorus
Phyllostachys decora
Sorbus decora
Thalictrum decorum

---

decorticans, decorticatus with peeling
  bark; stripped of bark
decrescens becoming less or narrower
decumanus very large

decumbens decumbent; prostrate but with upright tips

Cistus dansereaui 'Decumbens'

Correa decumbens

Cytisus decumbens

Hebe decumbens

---

decurrens, decursivus decurrent, running down the stem

Calocedrus decurrens

decurvatus, decurvus decurved

decussatus decussate

Melaleuca decussata

deflexus bent downward

Enkianthus deflexus

WHAT'S IN A NAME?

COLOUR: BLACK

anthracinus coal black, bluish black

ater pure black

atratus clothed in black for mourning; dark

carbonaceus black like charcoal

coracinus, corvinus crow black, shiny black

denigratus blackened

ebenaceus, ebenus black like ebony

mela-, melano- pure black

nigellus blackish

niger black

nigrescens, nigricans blackish, blackening

nigritus dressed in black

piceus brownish black, pitch black

pullus dusky, greyish black

deformis misshapen

dehiscens dehiscent

delapsus fallen away

---

delavayi, delavayanus named for the French botanist and missionary Jean Marie Delavay (1834–1895)

Asarum delavayi

Incarvillea delavayi

Iris delavayi

Ligustrum delavayanum

Magnolia delavayi

Osmanthus delavayi

Paeonia delavayi

Philadelphus delavayi

Podophyllum delavayi

---

delicatus delicate

deliciosus delicious

Actinidia deliciosa (kiwi fruit)

delphicus from Delphi, in Greece

deltoides triangular (like the Greek letter delta)

Dianthus deltoides

demissus hanging, drooping

dendro- tree-

dendroideus like a tree

dens-canis dog's tooth

Erythronium dens-canis

densiflorus densely flowered

Hedychium densiflorum

densus, densatus dense, compact

Tanacetum densum

CULTIVAR NAMES
Strange but True
D

PRIMULA AURICULA 'DAFTIE GREEN'

ARGYRANTHEMUM
DAISY CRAZY SERIES

IRIS SIBIRICA
'DANCE BALLERINA DANCE'

ALSTROEMERIA 'DANDY CANDY'

AEGOPODIUM PODAGRARIA
'DANGEROUS'

PELARGONIUM 'DARK SECRET'

IRIS 'DEMON'

PHYGELIUS × *RECTUS* 'DEVIL'S TEARS'

PHYSOCARPUS OPULIFOLIUS 'DIABOLO'

ORIGANUM DICTAMNUS
'DINGLE FAIRY'

PAEONIA LACTIFLORA 'DINNER PLATE'

PRIMULA AURICULA 'DIVINT DUNCH'

IRIS 'DIXIE PIXIE'

FUCHSIA 'DOCTOR FOSTER'

PHLOX PANICULATA 'DOGHOUSE PINK'

RHODODENDRON 'DOPEY'

RANUNCULUS FICARIA 'DOUBLE MUD'

HEDERA HELIX 'DUCKFOOT'

GENTIANA 'DUMPY'

PRIMULA AURICULA 'DUSTY MILLER'

COMMON DESCRIPTIVE TERMS

dentatus toothed
Dodecatheon dentatum
Hedera colchica 'Dentata'
Lavandula dentata
Ligularia dentata
Quercus dentata

DENTATUS

denticulatus with small teeth
  Primula denticulata
denudatus stripped (e.g. of leaves)
  Magnolia denudata
depauperatus impoverished, dwarfed
  Carex depauperata
dependens hanging down
deplanatus levelled
depressus flattened
  Juniperus communis 'Depressa
  Aurea'
diabolicus like a devil
  Acer diabolicum
dianthifolius with leaves like *Dianthus*
  Commelina dianthifolia
dicentrifolius with leaves like *Dicentra*
  Codonopsis dicentrifolia
dichotomus dividing repeatedly in two
  Eryngium dichotomum
dictyophyllus with leaves showing an
  obvious network of veins
  Berberis dictyophylla
didymus paired, two-lobed
  Monarda didyma
difformis unusual in form
  Vinca difformis

Daffodil *Narcissus*
Daisy *Bellis perennis*
Dame's violet *Hesperis matronalis*
Dandelion *Taraxacum*
Date palm *Phoenix dactylifera*
Dawn redwood *Metasequoia glyptostroboides*
Daylily *Hemerocallis*
Deadly nightshade *Atropa bella-donna*
Deadnettle *Lamium*
Deodar *Cedrus deodara*
Desert candle *Eremurus*

ELECAMPANE

Devil's walking stick *Aralia spinosa*
Dittany *Dictamnus albus*
Dock *Rumex*
Dog rose *Rosa canina*
Dog violet *Viola riviniana*
Dog's tooth violet *Erythronium dens-canis*
Dogwood *Cornus*
Douglas fir *Pseudotsuga menziesii*
Dove tree *Davidia involucrata*
Downy birch *Betula pubescens*
Dragon arum *Dracunculus vulgaris*
Dropwort *Filipendula vulgaris*
Durmast oak *Quercus petraea*
Dutchman's breeches *Dicentra spectabilis*
Dutchman's pipe *Aristolochia*
Dyer's greenweed *Genista tinctoria*
Eastern hemlock *Tsuga canadensis*
Eastern red cedar *Juniperus virginiana*
Eglantine *Rosa rubiginosa*
Egyptian bean *Lablab purpureus*
Elder *Sambucus*
Elecampane *Inula helenium*
Elephant's ears *Bergenia*
Elm *Ulmus*
English oak *Quercus robur*
European larch *Larix decidua*
Evening primrose *Oenothera*
Evergreen oak *Quercus ilex*
Everlasting pea *Lathyrus grandiflorus; Lathyrus latifolius*

diffractus broken in pieces
diffusus spreading
  Lithodora diffusa
digitatus hand-shaped; with fingers
  Carex digitata

digitiformis finger-shaped
digynus with two styles or carpels
  Sarcococca hookeriana var. *digyna*
dilatatus spread out, expanded
  Dryopteris dilatata

dilutus diluted, weak

diminutus small, diminished

dimorphus existing in two forms

COMMON DESCRIPTIVE TERMS

dioicus dioecious (i.e. with male and
female reproductive organs
on separate plants)
Aruncus dioicus
Carex dioica
Gymnocladus dioica
Silene dioica (red campion)

---

dioscoridis like *Dioscoria* (yam),
named for the 1st-century AD
Greek physician Dioscorides
Arum dioscoridis

diosmifolius, diosmatifolius with
leaves like *Diosma*
Hebe diosmifolia

diphyllus two-leaved
Cardamine diphylla

diplo- double-

dipsaceus, dipsacoides like *Dipsacus*
(teasel)
Carex dipsacea

dipterocarpus with two-winged
fruit

dipterus two-winged

directus straight

discedens dividing

disciformis disc-shaped

discolor of different colours
Aquilegia discolor

CULTIVAR NAMES
Colours: Green

CAMELLIA JAPONICA 'FOREST GREEN'

CONVALLARIA MAJALIS
'GREEN TAPESTRY'

ERYNGIUM YUCCIFOLIUM
'GREEN SWORD'

EUONYMUS JAPONICUS 'GREEN SPIDER'

HEBE 'GREENSLEEVES'

HEDERA HELIX 'GREEN RIPPLE'

HELLEBORUS FOETIDUS
'GREEN GIANT'

HEMEROCALLIS FULVA
'GREEN KWANSO'

HEMEROCALLIS 'GREEN FLUTTER'

HEUCHERA CYLINDRICA 'GREENFINCH'

HYDRANGEA PANICULATA
'GREENSPIRE'

IRIS 'GREEN SPOT'

JOVIBARBA HEUFFELII 'GREENSTONE'

JUNIPERUS CHINENSIS
'ROBUST GREEN'

KNIPHOFIA 'GREEN JADE'

MALUS DOMESTICA (APPLE)
'GREENSLEEVES'

NICOTIANA LANGSDORFFII
'LIME GREEN'

PACHYSANDRA TERMINALIS
'GREEN CARPET'

PITTOSPORUM EUGENIOIDES
'GREEN ELF'

PRIMULA AURICULA 'GREEN ISLE'

RANUNCULUS FICARIA 'GREEN PETAL'

SEDUM 'GREEN EXPECTATIONS'

SEMPERVIVUM 'GREEN APPLE'

ZANTEDESCHIA AETHIOPICA
'GREEN GODDESS'

Born in Scone, Scotland, David Douglas became a gardener and, by 1820, was working at the Glasgow Botanic Garden. Recruited by the Horticultural Society as a plant collector, in 1824 he left for the west coast of North America, sailing round Cape Horn and via the Galapagos Islands up to the Columbia River in Oregon. Over the next 10 years he brought many plants back to Britain, including penstemons, camassias and *Eschscholzia californica*, but he is best known for the conifers he introduced as seeds, including the Douglas fir, Monterey pine and Sitka spruce. Plants named for him include *Iris douglasiana*, *Limnanthes douglasii* and *Phlox douglasii*. Douglas died in Hawaii, aged only 35, killed by a bull in a terrible accident.

DAVID DOUGLAS
(1798–1834)

---

discretus separate

disjunctus separate

  Salvia disjuncta

dispersus, displicatus scattered

disruptus broken off

dissectus dissected; deeply divided

  Acer palmatum var. *dissectum*

dissimilis dissimilar

dissitus spaced out

dissolutus dissolved

distachyus with two spikes

  Ephedra distachya

distichophyllus with leaves in two rows

  Eustachys distichophylla

distichus in two rows

  Abeliophyllum distichum

distylus with two styles

  Acer distylum

diurnus diurnal, flowering by day

COMMON DESCRIPTIVE TERMS

divaricatus spreading, straggling

Aster divaricatus

Cotoneaster divaricatus

Phlox divaricata

Plagianthus divaricatus

---

divergens diverging

diversifolius with differing leaves

  Clematis × *diversifolia*

diversiformis of different forms

diversus opposite, distinct, turned in different directions

divisus divided

divulgatus widespread

divulsus torn apart

dodeca- 12- (e.g. *Dodecatheon*)

dodecandrus with 12 stamens

dolabratus shaped like a pickaxe

dolicho- long- (see below)

dolichopetalus with long petals
dolichostachyus with a long spike
  Carex dolichostachya
dolomiticus from the Dolomites
  Campanula dolomitica
domestica domesticated
  Malus domestica (apple)
  Nandina domestica
donax named after a kind of reed
  Arundo donax
douglasii, douglasianus (see panel
  opposite)
dracocephalus with a head like a
  dragon
  Fargesia dracocephala
dracunculus little dragon
  Artemisia dracunculus (tarragon)
drakensbergensis from the
  Drakensberg range in South
  Africa
  Scabiosa drakensbergensis
drummondii named for the Scottish
  plant-collecting brothers James
  (1786–1863) and Thomas
  Drummond (1793–1835)
  Phlox drummondii

COMMON DESCRIPTIVE TERMS

dulcis sweet, pleasant
Euphorbia dulcis
Hovenia dulcis
Lippia dulcis
Phyllostachys dulcis
Prunus dulcis (almond)

dumetorum of thickets and
  hedgerows
  Helleborus dumetorum
dumosus bushy
  Holodiscus dumosus
durandii named for the French
  nursery Durand Frères
  Clematis × *durandii*
dysentericus for the treatment
  of dysentery
  Pulicaria dysenterica

GENUS NAMES

D, E

Dahlia named for Dr Anders Dahl
(1751–1789), Swedish botanist
Davidia named for Abbé Jean Pierre
Armand David (see page 65)
Dicentra from Greek *di-*, two, and
kentron, spur
Dicksonia named for James Dickson
(1738–1822), British nurseryman
and botanist
Dierama in Greek, a funnel
Digitalis from Latin *digitus*, finger
Dodecatheon from Greek *dodeka*, 12,
and *theos*, god
Echinacea, Echinops from Greek
echinos, a hedgehog
Epilobium from Greek *epi*, on, and
lobos, pod
Equisetum from Latin *equus*, horse,
and *seta*, bristle
Erodium from Greek *erodios*, heron
Erythronium from Greek *erythros*, red
Eucomis from Greek *eu*, good, and
kome, hair

E

ebenaceus, ebenus black like ebony
ebracteatus without bracts
eburneus ivory white
   Eryngium eburneum
echinatus prickly, spiny
echiniformis shaped like a hedgehog
   or sea-urchin (*echinus*)
   Picea glauca 'Echiniformis'

CULTIVAR NAMES
Personal Names
E

CORNUS 'EDDIE'S WHITE WONDER'
CARDAMINE PRATENSIS 'EDITH'
CYRTANTHUS 'EDWINA'
PELARGONIUM 'EILEEN'
CLEMATIS 'ELEANOR'
FUCHSIA 'ELFRIDA'
POTENTILLA FRUTICOSA 'ELIZABETH'
ESCALLONIA LAEVIS 'GOLD ELLEN'
FUCHSIA 'ELSA'
GERANIUM SANGUINEUM 'ELSBETH'
MONARDA 'ELSIE'S LAVENDER'
GLADIOLUS 'ELVIRA'
IRIS 'BROADLEIGH EMILY'
DIASCIA 'EMMA'
ASTER ERICOIDES 'ESTHER'
PRIMULA 'EUGÉNIE'
ANTHEMIS TINCTORIA 'EVA'
PENSTEMON 'EVELYN'

---

echinocarpus with prickly fruit
echinocephalus with a prickly head
   Ptilostemon echinocephalus
ecorticatus (see also *excorticatus*)
   without bark
edgeworthii named for Michael
   Edgeworth (1812–1881), a British
   amateur botanist and plant collector
   Rhododendron edgeworthii

COMMON DESCRIPTIVE TERMS
edulis edible
Passiflora edulis
Canna edulis
Carpobrotus edulis
Phyllostachys edulis

---

effusus spreading, straggly
   Milium effusum
elaeagnifolius with leaves like
   Elaeagnus
   Pyrus elaeagnifolia
elasticus elastic
   Ficus elastica

COMMON DESCRIPTIVE TERMS
elatus tall; **elatior, elatius** taller
Aralia elata
Arrhenatherum elatius
Aspidistra elatior
Carex elata
Persicaria elata
Primula elatior
Rehmannia elata

WHAT'S IN A NAME?

PLANT LOOKALIKES

abietinus like *Abies* (fir)
acanthifolius with leaves like *Acanthus*
agavoides like *Agave*
alnifolius with leaves like *Alnus* (alder)
althaeoides like *Althaea* (hollyhock)
alyssoides like *Alyssum*
amaranthoides like *Amaranthus*
amygdaloides almond-like
anemoneflorus anemone-flowered
aquilegiifolius with leaves like *Aquilegia*
arundinaceus like a reed
bambusoides like bamboo
betulifolius with leaves like *Betula*
  (birch)
bryoides like moss
buxifolius box-leaved
calamus reedlike
cardunculus like a little thistle
clematideus like *Clematis*
cupressoides like *Cupressus* (cypress)
cyclamineus like *Cyclamen*
daphnoides like *Daphne*
dendroideus like a tree
dianthifolius with leaves like *Dianthus*
elaeagnifolius with leaves like *Elaeagnus*
ericoides like *Erica* (heath)
festucoides resembling *Festuca* (fescue)
ficifolius with leaves like *Ficus* (fig)
fuchsioides like *Fuchsia*
gentianoides like *Gentiana* (gentian)
gramineus like grass
graminifolius with grass-like leaves
ilicifolius with leaves like *Ilex* (holly)
iridiflorus with flowers like *Iris*
jasminoides like *Jasminum* (jasmine)
juncifolius with rushlike leaves
ligustrinus like *Ligustrum* (privet)
liliago lily-like

linifolius with leaves like *Linum* (flax)
malviflorus with flowers like *Malva*
  (mallow)
muscosus mossy, like moss
ocymoides like *Ocimum* (basil)
oleoides like *Olea* (olive)
opulifolius with maple-like leaves
orchidiflorus with flowers like orchid
pandanifolius with leaves like *Pandanus*
  (screw pine)
persicifolius with leaves like *Prunus
  persica* (peach tree)
persimilis very similar (e.g. to another
  species)
phlogopappus with seeds like *Phlox*
pinifolius with leaves like pine
platanoides like *Platanus*
plumbaginoides like *Plumbago*
prunifolia with leaves like *Prunus*
pseud-, pseudo- false (e.g.
  pseudoplatanus: 'false' *Platanus* or
  plane)
quercifolius with leaves like *Quercus*
  (oak)
ranunculoides like *Ranunculus*
rhamnoides like *Rhamnus* (buckthorn)
salicifolius with leaves like *Salix*
  (willow)
salviifolius with leaves like *Salvia*
scilloides like *Scilla*
taxifolius with leaves like *Taxus* (yew)
typhoides like *Typha* (bulrush)
ulmarius like *Ulmus* (elm)
urticifolius with leaves like *Urtica*
  (nettle)
uvarius like grapes
viburnoides like *Viburnum*
vitifolius with leaves like *Vitis* (vine)

COMMON DESCRIPTIVE TERMS

elegans elegant; *elegantissimus* very
elegant
Buxus sempervirens 'Elegantissima'
Clarkia elegans
Cornus alba 'Elegantissima'
Desmodium elegans
Echeveria elegans
Liatris elegans

elegantulus rather elegant
Acer elegantulum
eleuther- free, not joined
ellipsoidalis ellipsoid
Quercus ellipsoidalis
ellipticus elliptic
Garrya elliptica

elongatus elongated
Melianthus elongatus
elwesii named after the English
naturalist and arboriculturist Henry
John Elwes (1846–1922)
Galanthus elwesii
emeticus emetic, causing vomiting
eminens projecting
emodensis, emodi from the Himalayas
Paeonia emodi
empetriformis shaped like *Empetrium*
(crowberry)
Phyllodoce empetriformis
endressii named for Philip Endress
(1806–1831), a German
plant collector,
Geranium endressii
engelmannii named for George
Engelmann (1809–1884) a
German botanist and doctor living
and working in the USA
Picea engelmannii
enneaphyllus with nine leaflets
or leaves
Oxalis enneaphylla

COMMON DESCRIPTIVE TERMS

ensatus, ensiformis sword-shaped;
ensifolius sword-leaved
Aristea ensifolia
Inula ensifolia
Iris ensata
Juncus ensifolius
Kniphofia ensifolia

ephemerus ephemeral, short-lived
  Lysimachia ephemerum
epi- on, on top of
epilithicus (growing) on stones
epigaeus growing near the ground
equinus, equestris relating to horses

COMMON DESCRIPTIVE TERMS

erectus erect, upright
Hedera helix 'Erecta'
Potentilla erecta
Tagetes erecta
Trillium erectum

---

erianthus with woolly flowers
  Geranium erianthum
ericifolius with leaves like *Erica*
  (heath)
  Melaleuca ericifolia
ericoides like *Erica* (heath)
  Aster ericoides
ericsmithii named for Eric Smith
  (1917–1986), plantsman and
  gardener at Hadspen, Somerset
  Helleborus × *ericsmithii*
erinaceus like a hedgehog
  Dianthus erinaceus
eri-, erio- woolly- (see below)
eriocarpus with woolly fruits
  Potentilla eriocarpa
eriophyllus with woolly leaves
eriostemon with woolly stamens
erromenus vigorous, healthy
  Hosta undulata var. *erromena*

erubescens reddening, blushing
  Arisaema erubescens
erythro- red- (see below)
erythrocarpus with red fruits
  Lindera erythrocarpa

CULTIVAR NAMES
Foreign Expressions
E

ECLAIREUR (*French*)
SCOUT
(*PHLOX PANICULATA* 'ECLAIREUR')

ELFENAUGE (*German*)
ELF'S EYE
(*OMPHALODES VERNA* 'ELFENAUGE')

ELMFEUER (*German*)
ST ELMO'S FIRE
(*LOBELIA FULGENS* 'ELMFEUER')

ENZIANDOM (*German*)
GENTIAN DOME
(*HYDRANGEA MACROPHYLLA*
'ENZIANDOM')

ERDBLUT (*German*)
EARTH BLOOD
(*SEDUM SPURIUM* 'ERDBLUT')

ERSTE ZUNEIGUNG (*German*)
FIRST LOVE
(*PAPAVER ORIENTALE* 'ERSTE
ZUNEIGUNG')

ESCARBOUCLE (*French*)
CARBUNCLE
(*NYMPHAEA* 'ESCARBOUCLE')

ETOILE (*French*)
STAR
(*CLEMATIS* 'ETOILE ROSE',
CLEMATIS 'ETOILE VIOLETTE')

WHAT'S IN A NAME?

TEXTURE

acanth-, acantho- spiny, thorny
aculeatus prickly
adamantinus hard, steely
alutaceus leathery
amylaceus starchy
asper rough
asperatus roughened
asperrimus very rough
bombycifer with silky hairs
bullatus blistered, bubbled, studded
callosus with a hard skin
calvus bald, hairless
caperatus wrinkled
ceraceus waxy
cerefolius with waxy leaves
coactus felted
comatus, comosus tufted
coriaceus, coriarius leathery
corrugatus corrugated, furrowed
costatus ribbed
crinitus covered with long hairs
echinatus prickly, spiny
erio- woolly
eriocarpus with woolly fruits
farinosus mealy, covered with farina
fibrosus fibrous
flaccidus flaccid, soft
floccosus floccose, woolly
foveatus, foveolatus pitted
furfuraceus scurfy, covered with bran-
  like scales or powder
glaber glabrous (hairless, smooth)
glabellus rather smooth
glutinosus glutinous, sticky
gossypinus cottony
hirsutus hairy
hispidus bristly
laevigatus smooth

laevis soft
lanatus, laniger woolly
lanuginosus woolly
lisso- smooth
malacoides soft, supple
membranaceus like skin or membrane
mollis soft
mucosus slimy
muricatus rough with sharp points
nervosus fibrous, sinewy; with
  conspicuous veins
pilosus with long soft hairs, shaggy
plumarius, plumosus feathery
polytrichus with many hairs
pubens, pubescens, pubiger downy
pulverulentus dusty, powdery
pustulatus blistered or pimply
resinifer, resinosus resinous
rhytidophyllus with wrinkled leaves
rigens, rigidus stiff
rugosus, ruginosus, rugulosus wrinkled
scaber, scabrosus rough, gritty
sericeus silky
setaceus, setosus, setiferus bristly
spinosus, spinifer thorny
spinulifer with little spines
squamatus, squamosus scaly
squarrosus rough, scurfy; with
  protruding scales
stimulosus with stings or prickles
strigosus bristly
strobilaceus scaly like a pine-cone
subhirtellus rather hairy
tomentosus tomentose, woolly
velutinus velvety
verrucosus, verruculosus warty
villosus softly hairy
viscosus sticky, viscous

erythropodus with a red stem
  Alchemilla erythropoda
erythrosorus with red sori
  Dryopteris erythrosora
erythrostictus with red spots
  Sedum erythrostictum
esculentus edible
  Colocasia esculenta
esthonicus from Estonia
  Milium effusum var. *esthonicum*
estriatus not striped
etruscus from Tuscany, Italy
  Lonicera etrusca
eu- well, good (see below)
eudoxus of good repute
eucharis pleasing, agreeable
eugenioides like *Eugenia,* named for
  Prince Eugene of Savoy
  Pittosporum eugenioides

europaeus European
Asarum europaeum
Euonymus europaeus
Olea europaea
Tilia × *europaea*
Trollius europaeus

evenosus not conspicuously veined
  Hebe evenosa
evolutus unfolded
exaltatus raised high, tall
  Nephrolepis exaltata
exasperatus roughened

excelsus lofty, high, tall; *excelsior*
  higher
Doronicum × *excelsum*
Fraxinus excelsior
Ligustrum lucidum 'Excelsum
  Superbum'
Lobelia excelsa
Rhapis excelsa

excorticatus, ecorticatus without bark
  Fuchsia excorticata
exiguus small, weak
  Salix exigua
exiliflorus with small or slender
  flowers
  Liriope exiliflora
eximius distinguished, extraordinary
  Daphne cneorum 'Eximia'
exoniensis from Exeter (usually from
  the nursery of Veitch & Sons, see
  page 204)
  Passiflora × *exoniensis*
exoticus foreign
expansus spread out, expanded
  Juniperus chinensis 'Expansa
  Variegata'
exstipulatus exstipulate
  Passiflora exstipulatum
eystettensis named for the garden
  of Prince Bishop Conrad von
  Gemmingen at Eichstätt in
  Germany
  Narcissus 'Eystettensis'

F

fabiformis bean-shaped

facetus choice, fine
  Rhododendron facetum

faeroensis from the Faeroe Islands
  Alchemilla faeroensis

fagifolius with leaves like *Fagus* (beech)

falcatus falcate, curved like a sickle
  Polygonatum falcatum

fargesii named for the French missionary and plant hunter in China, Paul Farges (1844–1912)
  Abies fargesii

CULTIVAR NAMES
Personal Names
F

Rosa 'Felicia'

Jasminum officinale Fiona Sunrise ('Frojas')

Fragaria × *ananassa* (strawberry) 'Florence'

Erodium 'Fran's Delight'

Campanula persicifolia 'Frances'

Primula 'Francisca'

Pinus nigra 'Frank'

Clematis 'Frankie'

Fuchsia 'Fred's First'

Clematis 'Freda'

Rhododendron 'Freya'

---

farinaceus starchy
  Salvia farinacea

farinosus mealy, covered with farina
  Primula farinosa

farreri (see panel opposite)

fasciatus banded, striped

fasciculatus clustered, in bundles
  Primula fasciculata

fasciculiflorus with flowers in clusters
  Clematis fasciculiflora

COMMON DESCRIPTIVE TERMS

fastigiatus fastigiate; with an upright habit
  Gypsophila fastigiata
  Ilex crenata 'Fastigiata'
  Liriodendron tulipifera 'Fastigiatum'
  Malva alcea var. *fastigiata*
  Taxus baccata 'Fastigiata'

---

fastuosus proud, haughty
  Semiarundinaria fastuosa

faveolatus; favosus finely honeycombed

febrifugus driving out fever
  Dichroa febrifuga

fecundus fruitful; fertile

fenestralis with windowlike openings

fennicus Finnish

ferax fruit-bearing, fruitful, fertile

ferdinandi named for King Ferdinand of Bulgaria (1861–1948)
  Agave ferdinandi-regis
  Arabis ferdinandi-coburgi

ferens carrying, bearing
ferox fierce (generally used of very
  spiny plants)
  Ilex aquifolium 'Ferox Argentea'
  Pseudopanax ferox
ferreus iron grey; hard like iron
ferrugineus; ferruginosus rusty; light
  reddish brown
  Digitalis ferruginea
fertilis producing many seeds, fruitful
ferulaceus; ferulifolius like *Ferula*
  (giant fennel)
ferus wild
festalis, festivus festive, bright
  Hymenocallis × *festalis*
festucoides resembling *Festuca* (fescue)
fibrillosus fibrillose
fibrosus fibrous
  Dicksonia fibrosa
ficifolius with leaves like *Ficus carica*
  (fig)
figlinus terra cotta
figuratus formed, shaped
filamentosus, filarius with filaments
  or threads
  Yucca filamentosa
fili- thread (see below)
filicaulis with a threadlike stem
filici- fern- (*filix*; see below)
filicifolius with fernlike leaves
  Acacia filicifolia
filicoides fernlike
filiculoides like a small fern
  Azolla filiculoides

REGINALD FARRER (1880–1920)

The young Reginald Farrer took
a particular interest in the flora
of his native Yorkshire and went
on to explore the plant life of the
Alps, Japan and China, becoming
an outstanding plant collector
and author. On his travels in the
Far East he collected many species
that are now garden favourites,
such as *Buddleja alternifolia*, as
well as new species of primula and
rhododendron. He wrote extensively
about his travels, and about alpines:
his two-volume book *The English
Rock Garden* is a classic work on the
subject. Poor health always made
the danger and discomfort of a
plant hunter's life especially difficult
for Farrer, and he died aged only
40 on a trip to Burma, where he
is buried. Plants named after him
include *Viburnum farreri*, as well as
Gentiana farreri, which he found in
north-west China, and the alpine
Geranium farreri.

fimbriatus fimbriate, fringed
  Dianthus 'Fimbriatus'
firmus strong, firm
fissifolius with split leaves
fissilis, fissus split, cleft
fissuratus fissured
fistulosus fistular; hollow like a pipe,
  but closed at the ends
  Monarda fistulosa
flabellatus, flabelliformis fan-shaped
  Crataegus flabellata
flaccidus flaccid, soft
  Yucca flaccida
flagellaris whiplike
  Saxifraga flagellaris
flagellifer whip-bearing
  Carex flagellifera
flagelliformis shaped like a whip
flammeus; flammeolus flame-coloured
flammula small flame
  Clematis flammula

---

filifer thread-bearing
  Washingtonia filifera
filifolius thread-leaved
filiformis threadlike
  Juncus filiformis 'Spiralis'
filipendulinus like *Filipendula*
  (meadowsweet)
  Achillea filipendulina
filipendulus hanging by a thread
filipes with a threadlike foot or stalk
  Rosa filipes
filix fern
  Athyrium filix-femina (lady fern)

COMMON DESCRIPTIVE TERMS

flavus pure yellow; **flavens** yellow;
  flavescens, flavicans, flavidus
  yellowish
  Chionochloa flavicans
  Crocus flavus subsp. *flavus*
  Disporum flavens
  Glaucium flavum
  Ilex aquifolium 'Flavescens'
  Linum flavum
  Pyracantha rogersiana 'Flava'
  Rhododendron flavidum

flavissimus very yellow

flavovirens yellowish green

fletcheri named for Harold Fletcher (1907–1978), British botanist, Director of RHS Garden Wisley, Surrey, and Keeper of Royal Botanic Garden Edinburgh
Chamaecyparis lawsoniana 'Fletcheri'

flexibilis; flexilis flexible, pliant
Pinus flexilis

flexipes with a flexible stem or foot
Trillium flexipes

COMMON DESCRIPTIVE TERMS
flexuosus zigzag
Corydalis flexuosa
Deschampsia flexuosa
Helenium flexuosum
Phyllostachys flexuosa

floccosus floccose, woolly
Cotoneaster floccosus

floralis floral, relating to the flower

florentinus from Florence, Italy
Cistus × *florentinus*

COMMON DESCRIPTIVE TERMS
flore pleno with double flowers
Clematis florida var. *flore-pleno*
Galanthus nivalis f. *pleniflorus* 'Flore Pleno'
Genista tinctoria 'Flore Pleno'
Ranunculus ficaria flore-pleno

WHAT'S IN A NAME?
COLOUR CLUES: GENERAL COLOUR TERMS

ater, atro- dark, blackish

bicolor bicoloured

concolor the same colour all over; of the same colour

discolor of different colours

floridus bright

fucatus painted, coloured

fulgens, fulgidus shining, brightly coloured

fuscopictus dark-coloured

fuscotinctus dark-tinged

fuscus dark

impolitus matt

iridescens iridescent

laetus bright

lucidus shining, clear, transparent

margaritaceus pearly

mutabilis changing, changeable (e.g. in colour)

nitens, nitidus shining

obscurus obscure, indistinct, dark

opacus shady, dark, opaque

pallescens rather pale, becoming pale

pallidus pale

pallidulus rather pale

phaeus dusky

pictus (literally) painted; vividly coloured

polychromus many-coloured

pullus dark in colour

purus pure

refulgens shining

splendens brilliant

squalidus dirty

tricolor three-coloured

vernicosus varnished

versicolor in various colours

Fairy moss *Azolla filiculoides*
False acacia *Robinia pseudoacacia*
False cypress *Chamaecyparis*
False indigo *Baptisia australis*
Fawn lily *Erythronium revolutum*
Feather grass *Stipa*
Fennel *Foeniculum vulgare*
Fern-leaved beech *Fagus sylvatica* var. *heterophylla* 'Aspleniifolia'
Fescue *Festuca*
Feverfew *Tanacetum parthenium*
Field maple *Acer campestre*
Fig *Ficus carica*
Figwort *Scrophularia*
Fir *Abies*
Firethorn *Pyracantha*
Fireweed *Chamerion angustifolium*
Fishbone cotoneaster *Cotoneaster horizontalis*
Flag *Iris*
Flame of the forest *Pieris*; *Butea monosperma*

Flannel bush *Fremontodendron californicum*
Flax *Linum*
Fleabane *Erigeron*
Flowering currant *Ribes sanguineum*
Flowering rush *Butomus umbellatus*
Foam flower *Tiarella cordifolia*
Foam of May *Spiraea* 'Arguta'
Forget-me-not *Myosotis*
Fountain grass *Pennisetum alopecuroides*
Four o'clock flower *Mirabilis jalapa*
Foxglove *Digitalis*
Foxglove tree *Paulownia tomentosa*
Foxtail lily *Eremurus*
French lavender *Lavandula stoechas*
French marigold *Tagetes patula*
Fringe tree *Chionanthus*
Fritillary *Fritillaria*
(Fruits, see page 169)
Fuji cherry *Prunus incisa*
Fumitory, Fumeroot *Corydalis*
Furze *Ulex*

FEVERFEW

floribundus profusely flowering
  Rhododendron floribundum
floridanus from Florida, USA
  Illicium floridanum
floridulus rather free-flowering
  Miscanthus floridulus

<small>COMMON DESCRIPTIVE TERMS</small>
floridus flowering abundantly
Clematis florida
Cornus florida
Primula florida
Weigela florida

---

florifer bearing flowers, flowering
florindae named by the British plant
  hunter Frank Kingdon Ward (see
  page 210) for his first wife Florinda
  Primula florindae
-florus -flowered (e.g. *parviflorus*,
  small-flowered)
flos flower
flos-cuculi flower of the cuckoo
  Lychnis flos-cuculi (ragged robin)
flos-jovis flower of Jove
  Lychnis flos-jovis
fluminalis, flumineus of rivers
fluminensis from Rio de Janeiro
  Tradescantia fluminensis
fluviatilis of rivers
  Equisetum fluviatile
foemineus female
foeniculaceum resembling *Foeniculum*
  (fennel)

GENUS NAMES
F

Festuca in Latin, a stalk or straw
Forsythia named for William
  Forsyth (1737–1804), Scottish
  horticulturist
Fothergilla named for Dr John
  Fothergill (1712–1780), English
  doctor and expert on American
  plants
Fritillaria from Latin *fritillus*,
  a dice-box
Fuchsia named for Leonhart Fuchs
  (1501–1566), German professor
  and herbalist

---

foetidissimus very evil-smelling
  Iris foetidissima
foetidus stinking, evil-smelling
  Helleborus foetidus
foliaceus leafy, leaf-like
foliatus leaf-bearing
foliosissimus very leafy
  Polemonium foliosissimum
foliosus leafy, many-leaved
  Kniphofia foliosa
-folius -leaved (e.g. *grandifolius*,
  large-leaved)
follicularis follicle-like
fontanesianus, fontanesii named for
  the French botanist and author
  René Desfontaines (1750–1833)
  Leucothoe fontanesiana
fontanus; fontinalis growing in or
  near springs

foratus pierced with holes
forficatus shaped like scissors
formicarius attractive to ants; of ants
formis -shaped (e.g. *cruciformis*)

formosanus from Taiwan (Formosa)
Lilium formosanum
Liquidambar formosana
Ophiopogon formosanus
Pleione formosana
Tricyrtis formosana

formosissimus very beautiful
formosus finely formed, handsome,
beautiful
Leycesteria formosa
fornicalis, fornicatus arched
forrestii (see panel below)
fortis strong, vigorous

fortunei named for the Scottish
plant collector Robert Fortune
(1812–1880)
Euonymus fortunei
Eupatorium fortunei
Hosta fortunei
Lysimachia fortunei
Mahonia fortunei
Paulownia fortunei
Rhododendron fortunei
Trachycarpus fortunei

foveatus pitted
foveolatus rather pitted
fractiflexus zigzag
fractus broken
fragarioides like *Fragaria* (strawberry)
fragilis fragile
Salix fragilis

**GEORGE FORREST
(1873–1932)**

Born in Falkirk, Scotland, George Forrest travelled to Australia as a young man, returning to work in the Botanic Garden in Edinburgh. This proved to be good grounding for his later career: his first trip, to China in 1904, was followed by six more expeditions. A courageous and prolific plant collector, he braved many dangers in order to send back carefully classified specimens and seeds of thousands of plants, aided by teams of Chinese villagers. Rhododendrons, including *Rhododendron forrestii*, were his speciality, but other discoveries included the very special *Gentiana sino-ornata*. Perhaps the most familiar of the plants named for Forrest is *Pieris formosa* var. *forrestii*; others include *Primula forrestii*, *Iris forrestii* and *Hypericum forrestii*.

FASAN (*German*) PHEASANT
(*HYDRANGEA MACROPHYLLA* 'FASAN')

FÉLICITÉ PERPÉTUE (*French*)
LASTING HAPPINESS
(*ROSA* 'FÉLICITÉ PERPÉTUE')

FERNER OSTEN (*German*) FAR EAST
(*MISCANTHUS SINENSIS* 'FERNER OSTEN')

FEUER (*German*) FIRE
(*ASTILBE* 'FEUER')

FEUERKERZE (*German*)
FIRE CANDLE
(*LYTHRUM SALICARIA* 'FEUERKERZE')

FEUERMEER (*German*)
SEA OF FIRE
(*GEUM* 'FEUERMEER')

FLAMMENSPIEL (*German*)
DANCING FLAMES
(*HELENIUM* 'FLAMMENSPIEL')

FLOCON DE NEIGE (*French*)
SNOWFLAKE
(*FUCHSIA* 'FLOCON DE NEIGE')

FRAÎCHE BEAUTÉ (*French*)
COOL BEAUTY
(*PELARGONIUM* 'FRAÎCHE BEAUTÉ')

FRÜHLINGSGOLD (*German*)
SPRING GOLD
(*ROSA* 'FRÜHLINGSGOLD')

FRÜHLINGSHIMMEL (*German*)
SPRING SKY
(*PULMONARIA SACCHARATA*
'FRÜHLINGSHIMMEL')

FRÜHLINGSZAUBER (*German*)
SPRING MAGIC
(*ARABIS BLEPHAROPHYLLA*
'FRÜHLINGSZAUBER')

FYRVERKERI (*Swedish*) FIREWORKS
(*OENOTHERA FRUTICOSA* 'FYRVERKERI')

fragrans fragrant
   Petasites fragrans
fragrantissimus very fragrant
   Lonicera fragrantissima
frangulus brittle
   Rhamnus frangula
fraseri named for the London
   nurseryman and plant collector
   John Fraser (1750–1811)
   Photinia × *fraseri*
fraxinifolius with leaves like
   Fraxinus (ash)
   Pterocarya fraxinifolia

frigidus cold; growing in cold places
   Cotoneaster frigidus
frikartii named for Carl Frikart
   (1879–1964), a Swiss nurseryman
   Aster × *frikartii*
frondosus leafy, leaf-like, leaf-bearing
   Primula frondosa
fructifer, frugifer fruit-bearing
fructu albo with white fruits
   Iris foetidissima 'Fructu Albo'
fructu luteo with yellow fruits
   Ilex aquifolium 'Pyramidalis Fructu
   Luteo'

CULTIVAR NAMES
Colours: Blue

BUDDLEJA DAVIDII 'EMPIRE BLUE'

CAMPANULA COCHLEARIFOLIA 'BAVARIA BLUE'

CARYOPTERIS × *CLANDONENSIS* 'HEAVENLY BLUE'

CEANOTHUS 'BLUE MOUND'

CHAMAECYPARIS LAWSONIANA 'PEMBURY BLUE'

CORYDALIS FLEXUOSA 'BLUE PANDA'

CROCUS CHRYSANTHUS 'BLUE PEARL'

DELPHINIUM 'BLUE DAWN'

ECHINOPS BANNATICUS 'TAPLOW BLUE'

ERYNGIUM ALPINUM 'BLUE STAR'

FESTUCA GLAUCA 'ELIJAH BLUE'

GENTIANA 'BLUE SILK'

GERANIUM 'JOHNSON'S BLUE'

HOSTA 'BLUE ANGEL'

HYACINTHUS ORIENTALIS 'DELFT BLUE'

HYDRANGEA SERRATA 'BLUEBIRD'

IPOMOEA TRICOLOR 'HEAVENLY BLUE'

IRIS 'BLUE DENIM'

JUNIPERUS HORIZONTALIS 'BLUE CHIP'

LAVANDULA ANGUSTIFOLIA 'LODDON BLUE'

PEROVSKIA 'BLUE SPIRE'

PRUNUS DOMESTICA (PLUM) 'BLUE TIT'

RHODODENDRON 'BLUE DANUBE'

RUTA GRAVEOLENS 'JACKMAN'S BLUE'

SALVIA PATENS 'CAMBRIDGE BLUE'

VERONICA PEDUNCULARIS 'GEORGIA BLUE'

VIOLA 'BELMONT BLUE'

COMMON DESCRIPTIVE TERMS

fruticosus, frutescens, fruticans
shrubby

Argyranthemum frutescens

Jasminum fruticans

Oenothera fruticosa

Penstemon fruticosus

Phlomis fruticosa

Potentilla fruticosa

Teucrium fruticans

---

fruticulosus shrubby and small
  Matthiola fruticulosa
fucatus painted, coloured
fuchsii named after the German
  physician Leonhart Fuchs (1501–
  1566), a German professor and
  herbalist
  Dactylorhiza fuchsii
fuchsioides like *Fuchsia*
  Begonia fuchsioides
fuegianus from Tierra del Fuego,
  Chile/Argentina
fugax fleeting, transitory, ephemeral
  Moraea fugax

COMMON DESCRIPTIVE TERMS

fulgens, fulgidus shining, brightly
  coloured

Fuchsia fulgens

Lobelia fulgens

Pelargonium fulgidum

Rudbeckia fulgida

Salvia fulgens

fuligineus; fuliginosus dark brown, sooty

fullonum of fullers
   Dipsacus fullonum (teasel)

fulmineus like lightning

COMMON DESCRIPTIVE TERMS

fulvus, fulvellus, fulvescens, fulvidus
tawny yellow, yellowish brown
   Digitalis × *fulva*
   Hemerocallis fulva
   Rhododendron fulvum

DIPSACUS FULLONUM

fumariifolius with leaves like *Fumaria* (fumitory)
   Corydalis fumariifolia

fumeus, fumidus, fumosus smoky, grey-brown

funalis rope-like

funebris funereal; of cemeteries

fungiformis mushroom-shaped

funicularis, funiculatus rope-like

funiculosus arranged in ropes or bundles

furcatus forked; with two long lobes

furfuraceus scurfy, covered with bran-like scales or powder

fuscopictus dark-coloured
   Disporopsis fuscopicta

fuscoruber dark red
   Corylus avellana 'Fuscorubra'

fuscotinctus dark-tinged
   Crocus chrysanthus var. *fuscotinctus*

fuscus dark; blackish brown

fusiformis fusiform, shaped like a spindle, thicker in the middle than at the ends

WHAT'S IN A NAME?

COUNTRIES OF EUROPE

anglicus England
austriacus Austria
belgicus Belgium or the Netherlands
bulgaricus Bulgaria
cambricus Wales
croaticus Croatia
gallicus France
germanicus Germany
graecus Greece
hispanicus Spain
italicus Italy
lusitanicus Portugal
scoticus Scotland
suecicus Sweden

G

gaditanus from Cadiz in Spain
  Narcissus gaditanus
gala- milk-
  Galactites; Galanthus
galeatus with a helmet or helmet-like
  covering
  Arisaema galeatum
gallicus Gallic, from France
  Rosa gallica

galpinii named for Ernest Galpin
  (1858–1941), a South African
  banker and plant collector
  Kniphofia galpinii
gardenii named for Dr Alexander
  Garden (1730–1791)
  Fothergilla gardenii
geminiflorus with twin flowers
  Polygonatum geminiflorum
gemmifer bearing buds
  Primula gemmifera
gemmosus jewelled
  Crataegus gemmosa

CULTIVAR NAMES
Colours: Gold

AUCUBA JAPONICA 'GOLDEN KING'

BERBERIS THUNBERGII BONANZA GOLD
  ('BOGOZAM')

CALLUNA VULGARIS 'GOLD HAZE'

CAREX OSHIMENSIS 'EVERGOLD'

CHAMAECYPARIS LAWSONIANA
  'ELLWOOD'S GOLD'

COTINUS GOLDEN SPIRIT ('ANCOT')

CYTISUS × PRAECOX 'ALLGOLD'

ELAEAGNUS × EBBINGEI 'COASTAL GOLD'

ERICA ARBOREA 'ALBERT'S GOLD'

EUONYMUS FORTUNEI 'EMERALD 'N'
  GOLD'

EUPHORBIA CHARACIAS SUBSP. WULFENII
  'LAMBROOK GOLD'

FORSYTHIA 'GOLDEN TIMES'

HEDERA HELIX 'GOLDCHILD'

HEMEROCALLIS 'GOLDEN CHIMES'

HOSTA 'GOLD STANDARD'

ILEX × ALTACLERENSIS 'GOLDEN KING'

ILEX CRENATA 'GOLDEN GEM'

JUNIPERUS × PFITZERIANA 'OLD GOLD'

LIGULARIA 'GREGYNOG GOLD'

LUMA APICULATA 'GLANLEAM GOLD'

MALUS DOMESTICA (APPLE) 'JONAGOLD'

MALUS × ZUMI 'GOLDEN HORNET'

MELISSA OFFICINALIS 'ALL GOLD'

NARCISSUS 'GOLDFINGER'

ORIGANUM VULGARE 'GOLD TIP'

PHYSOCARPUS OPULIFOLIUS 'DART'S
  GOLD'

PITTOSPORUM TENUIFOLIUM 'WARNHAM
  GOLD'

PYRACANTHA 'GOLDEN CHARMER'

RHODODENDRON 'GOLDEN TORCH'

RUBUS COCKBURNIANUS 'GOLDENVALE'

SAMBUCUS RACEMOSA 'SUTHERLAND
  GOLD'

generosus noble, eminent
  Populus × *generosa* 'Beaupré'
gentianoides like *Gentiana* (gentian)
  Veronica gentianoides
georgianus from Georgia (USA)
  Quercus georgiana
georgicus from Georgia (Caucasus)
  Pulsatilla georgica
geraniifolius with leaves like
  Geranium
  Androsace geraniifolia
germanicus from Germany
  Iris germanica
gibbosus hunchbacked; swollen
  on one side
  Patrinia gibbosa
gibraltaricus from Gibraltar
  Iberis gibraltarica

COMMON DESCRIPTIVE TERMS

giganteus giant, gigantic, very large
Allium giganteum
Cardiocrinum giganteum
Eryngium giganteum
Miscanthus × *giganteus*
Sequoiadendron giganteum
Stipa gigantea

---

gigas giant
  Angelica gigas
giraldianus, giraldii named for
  Giuseppe Giraldi (1848–1901), an
  Italian missionary and plant hunter
  Callicarpa bodinieri subsp. *giraldii*

CULTIVAR NAMES
Personal Names
G

CLEMATIS 'GABRIELLE'
OSTEOSPERMUM 'GEMMA'
CLEMATIS 'GEORG'
IRIS 'GEORGE'
ROSA GERALDINE ('PEAHAZE')
IRIS 'GORDON'
COTINUS 'GRACE'
ERICA × *WATSONII* 'GWEN'
DAHLIA 'GWYNETH'

---

glabellus rather glabrous
  Epilobium glabellum
glaber glabrous (hairless, smooth)
  Penstemon glaber
glabratus almost glabrous
  Hoheria glabrata

GENUS NAMES
G

Galanthus from Greek *gala*, milk,
  and *anthos*, flower
Gaura from Greek *gauros*, superb
Gazania named for Theodore of
  Gaza (1398–1478), translator of the
  botanical writings of Theophrastus
Geranium from Greek *geranos*, crane
Gladiolus in Latin, a small sword
Glyceria from Greek *glykys*, sweet
Gunnera named for Johan Gunnerus,
  Norwegian botanist and bishop
Gypsophila from Latin *gypsos*, chalk
  or gypsum, and *philos*, loving

glabrescens becoming glabrous
  Corylopsis glabrescens
glacialis from glaciers or cold places
  Artemisia glacialis

COMMON DESCRIPTIVE TERMS
glandulosus glandular
Arctostaphylos glandulosa
Calamintha nepeta subsp. *glandulosa*
Erodium glandulosum
Prunus glandulosa

---

glaucescens rather glaucous, becoming glaucous
  Primula glaucescens
glaucophyllus with glaucous leaves
  Hebe glaucophylla

COMMON DESCRIPTIVE TERMS
glaucus glaucous
Canna glauca
Carex glauca
Coronilla valentina subsp. *glauca*
Erigeron glaucus
Euphorbia glauca
Festuca glauca
Phyllostachys glauca
Picea glauca
Pinus pumila 'Glauca'
Rosa glauca

---

globiferus bearing globe-shaped clusters
  Androsace globifera
globispica with globose spikes
  Betula globispica

COMMON DESCRIPTIVE TERMS
globosus globular, spherical
Acer platanoides 'Globosum'
Buddleja globosa
Magnolia globosa
Picea pungens 'Globosa'

---

globulifer bearing small globe-shaped clusters

Galingale *Cyperus*
Gardener's garters *Phalaris arundinacea*
  var. *picta*
Gas plant *Dictamnus albus*
Gayfeather *Liatris spicata*
Gean *Prunus avium*
Gentian *Gentiana*
Geranium (tender) *Pelargonium*
Germander *Teucrium*
Germander speedwell *Veronica*
  chamaedrys
Giant fennel *Ferula communis*
Giant hogweed *Heracleum*
  mantegazzianum
Gillyflower *Matthiola*
Ginger, Ginger lily *Hedychium;*
  Zingiber
Glasswort *Salicornia europaea*
Globe amaranth *Gomphrena globosa*
Globe flower *Trollius*
Globe thistle *Echinops ritro*
Glory bush *Tibouchina*
Glory flower *Clerodendrum bungei;*
  Eccremocarpus scaber
Glory of the snow *Chionodoxa*
Goat willow *Salix caprea*
Goat's beard *Aruncus dioicus*
Goat's rue *Galega*
Golden oak of Cyprus *Quercus alnifolia*
Golden oats *Stipa gigantea*
Golden rain *Laburnum*
Golden rain tree *Koelreuteria paniculata*
Golden rod *Solidago*
Goldilocks *Ranunculus auricomus*
Gorse *Ulex*
Granny's bonnets *Aquilegia vulgaris*
Grape hyacinth *Muscari*
Grape vine *Vitis vinifera*

GLOBE FLOWER

Grass of Parnassus *Parnassia palustris*
Great mullein *Verbascum thapsus*
Great white cherry *Prunus* 'Taihaku'
Grey alder *Alnus incana*
Grey birch *Betula populifolia*
Grey poplar *Populus canescens*
Gromwell *Lithospermum officinale;*
  Buglossoides
Ground elder *Aegopodium podagraria*
Ground ivy *Glechoma hederacea*
Groundsel *Senecio vulgaris*
Guelder rose *Viburnum opulus*
Gum tree *Eucalyptus*

globulus like a little ball
  Eucalyptus globulus
glomeratus clustered in a head
  Campanula glomerata
gloriosus glorious
  Yucca gloriosa
glutinosus glutinous, sticky
  Eucryphia glutinosa
glycyrrhiza liquorice
  Polypodium glycyrrhiza

gmelinii named for a German
  naturalist, Johann Gmelin (1709–
  1755)
  Artemisia gmelinii
gossypinus like cotton
  Gerbera gossypina
gracilifolius with slender leaves
  Clematis gracilifolia
gracilipes with a slender stalk
  Primula gracilipes

COMMON DESCRIPTIVE TERMS
gracilis slender
Deutzia gracilis
Fuchsia magellanica var. *gracilis*
Galanthus gracilis
Geranium gracile
Mentha × *gracilis* (ginger mint)

---

gracillimus very slender
  Miscanthus sinensis 'Gracillimus'
graecus Greek
  Cyclamen graecum
gramineus like grass
  Iris graminea
graminifolius with grass-like leaves
  Polygonatum graminifolium
granatus with many seeds
  Punica granatum (pomegranate)
grandiceps with a large head
  Dryopteris filix-mas 'Grandiceps
  Wills'
grandidentatus with large teeth
  Fagus sylvatica 'Grandidentata'

CAMPANULA GLOMERATA

CULTIVAR NAMES
Foreign Expressions
G

GARTENSONNE (*German*)
GARDEN SUN
(*HELENIUM* 'GARTENSONNE')

GELBE MANTEL (*German*)
YELLOW CLOAK
(*IRIS* 'GELBE MANTEL')

GERBE D'OR (*French*)
GOLD SPRAY
(*CROCOSMIA* 'GERBE D'OR')

GEWITTERWOLKE (*German*)
THUNDERCLOUD
(*MISCANTHUS SINENSIS*
'GEWITTERWOLKE')

GLOCKENTURM (*German*)
BELL TOWER
(*BERGENIA* 'GLOCKENTURM')

GLOIRE DE ... (*French*)
GLORY OF ...
(*CEANOTHUS* × *DELILEANUS* 'GLOIRE
DE VERSAILLES'; *HEDERA CANARIENSIS*
'GLOIRE DE MARENGO')

GLUT (*German*)
GLOW
(*ASTILBE* 'GLUT')

GOLDBUKETT (*Swedish*)
GOLDEN BOUQUET
(*RHODODENDRON* 'GOLDBUKETT')

GOLDENE JUGEND (*German*)
GOLDEN YOUTH
(*HELENIUM* 'GOLDENE JUGEND')

GOLDFUCHS (*German*)
GOLD FOX
(*HELENIUM* 'GOLDFUCHS')

GOLDGEFIEDER (*German*)
GOLD PLUME
(*HELIOPSIS HELIANTHOIDES* VAR.
SCABRA 'GOLDGEFIEDER')

GOLDKLUMPEN (*German*)
GOLDEN CLOGS
(*POTENTILLA AUREA* 'GOLDKLUMPEN')

GOLDKRONE (*German*)
GOLDEN CROWN
(*RHODODENDRON* 'GOLDKRONE')

GOLDQUELLE (*German*)
GOLD FOUNTAIN
(*RUDBECKIA LACINIATA* 'GOLDQUELLE')

GOLDRAUSCH (*German*)
GOLD RUSH
(*HELENIUM* 'GOLDRAUSCH')

GOLDSTURM (*German*)
GOLD STORM
(*RUDBECKIA FULGIDA* VAR. *SULLIVANTII*
'GOLDSTURM')

GOLDTAU (*German*) GOLD DEW
(*DESCHAMPSIA CESPITOSA* 'GOLDTAU')

GOSHIKI (*Japanese*)
FIVE-COLOURED
(*OSMANTHUS HETEROPHYLLUS*
'GOSHIKI')

GRAUE WITWE (*German*)
GREY WIDOW
(*PAPAVER ORIENTALE* 'GRAUE WITWE')

GROSSE FONTÄNE (*German*)
BIG FOUNTAIN
(*MISCANTHUS SINENSIS*
'GROSSE FONTÄNE')

COMMON DESCRIPTIVE TERMS

grandiflorus large-flowered
Abelia × *grandiflora*
Clematis montana 'Grandiflora'
Digitalis grandiflora
Erythronium grandiflorum
Hydrangea anomala 'Grandiflora'
Magnolia grandiflora
Tellima grandiflora
Trillium grandiflorum

grandifolius with large leaves
  Achillea grandifolia
grandis big

gratianopolitanus from Grenoble in France
  Dianthus gratianopolitanus (Cheddar pink)

COMMON DESCRIPTIVE TERMS

graveolens strongly scented
Anethum graveolens (dill)
Apium graveolens
Pelargonium graveolens
Ruta graveolens (rue)

grayi named for the American botanist Asa Gray (1810–1888)
  Carex grayi
gregarius gregarious
  Nematanthus gregarius
griffithii named for William Griffith (1810–1845), a British botanist
  Euphorbia griffithii
griseus grey
  Acer griseum
groenlandicus from Greenland
  Ledum groenlandicum
grossularioides like gooseberry
  Pelargonium grossularioides
gruinus like a crane
  Erodium gruinum
gunnii named for Ronald Gunn (1808–1881), a South African-born Tasmanian botanist
  Eucalyptus gunnii
guttatus spotted
  Helleborus orientalis subsp. *guttatus*

CULTIVAR NAMES
Say it with Flowers

Galanthus nivalis 'April Fool'

Iris sibirica 'Baby Sister'

Narcissus 'Best of Luck'

Rosa Best Wishes ('Chessnut')

Hosta 'Big Daddy'

Malva sylvestris 'Brave Heart'

Rosa Congratulations ('Korlift')

Hemerocallis 'Cool It'

Rosa Crazy for You ('Wekroalt')

Dianthus 'Dad's Favourite'

Iris ensata 'Darling'

Rhododendron 'Dear Grandad'

Rosa 'Dearest'

Paeonia lactiflora 'Do Tell'

Fuchsia 'Forget-me-Not'

Alstroemeria 'Friendship'

Trifolium repens 'Good Luck'

Rheum × hybridum (rhubarb) 'Grandad's Favourite'

Dianthus 'Gran's Favourite'

Pelargonium 'Greetings'

Liquidambar styraciflua 'Happidaze'

Rosa Happy Retirement ('Tantoras')

Tulipa 'Heart's Delight'

Hosta 'Hope'

Lilium 'Journey's End'

Rosa Keep in Touch ('Hardrama')

Fuchsia 'Keepsake'

Rosa Keepsake ('Kormalda')

Iris 'Let's Elope'

Rosa Love Knot ('Chewglorious')

Primula auricula 'Lovebird'

Rosa Loving Memory ('Korgund')

Rosa Marry Me ('Dicwonder')

Lavatera × clementii Memories ('Stelav')

Malus domestica (apple) 'Mother'

Paeonia lactiflora 'Mother's Choice'

Dahlia 'My Valentine'

Dahlia 'New Baby'

Hosta 'Remember Me'

Fuchsia 'Remembrance'

Rhododendron 'Second Honeymoon'

Lilium 'Silly Girl'

Fuchsia 'Sister Sister'

Pelargonium 'Something Special'

Pelargonium 'Souvenir'

Rosa Special Child ('Tanaripsa')

Nepeta subsessilis 'Sweet Dreams'

Tulipa 'Sweetheart'

Rosa Thank You ('Chesdeep')

Fuchsia 'Three Cheers'

Rosa Top Marks ('Fryministar')

Rosa Warm Wishes ('Fryxotic')

Iris sibirica 'Welcome Return'

Rosa 'Well Done'

Camellia japonica 'Yours Truly'

See also Anniversaries (page 183); Birthdays (page 150); Christmas (page 98); Special Occasions; (page 107); Weddings (page 188).

H

haageanus named for a German
  seedsman, J. Haage (1826–1878)
  Lychnis haageana
haastii named for Sir Johann von
  Haast (1824–1887), a German
  plant collector in New Zealand
  Olearia × haastii
habrotrichus with soft hairs
hadriaticus from the Adriatic
haem- blood-red (see below)
haemanthus with blood-red flowers
haematocalyx with a blood-red calyx
  Dianthus haematocalyx
haematocarpus with blood-red fruits
haematodes blood-red
  Rhododendron haematodes
hakusanensis from Mount Haku
  in Japan
  Sanguisorba hakusanensis
halo- salt
halophilus salt-loving
hamatus, hamosus hooked, barbed
hamulosus with small hooks
haplo- single (see below)
haplocalyx with a single calyx
  Mentha haplocalyx
haplocaulis with a single stem
haplostichus with a single row
harpophyllus with sickle-shaped
  leaves

hastatus with a spear; spear-shaped
  Verbena hastata
hastifolius with spear-shaped leaves
  Scutellaria hastifolia
hebe- downy (see below)
hebecarpus with downy fruit
hebepetalus with downy petals
  Indigofera hebepetala
hebephyllus with downy leaves
hebetatus made dull or blunt
hederaceus like *Hedera* (ivy)
  Glechoma hederacea
hederifolius with leaves like *Hedera*
  Cyclamen hederifolium
heldreichii named for a German
  botanist, Theodor von Heldreich
  (1822–1902)
  Pinus heldreichii
heleo-, helo- marsh-
helianthoides like *Helianthus*
  Heliopsis helianthoides
helicus coiled
heli-, helio- sun-
heliolepis with glittering scales
helix in a spiral; twining
  Hedera helix
hellenicus Greek
helodes of marshes

COMMON DESCRIPTIVE TERMS

helveticus Swiss
Erysimum helveticum
Onosma helvetica
Salix helvetica

---

helvolus pale yellow
   Nymphaea 'Pygmaea Helvola'
helvus light bay in colour
hemi- half-
hemisphaericus hemispherical
hemitrichotus hairy on one side only
hemsleyanus, hemsleyi named for
   William Hemsley (1843–1924),
   a gardener and botanist at Kew
   Aconitum hemsleyanum
   Sorbus hemsleyi

COMMON DESCRIPTIVE TERMS

henryanus, henryi named for the
   Irish botanist and plant collector
   Augustine Henry (1857–1930)
Acer henryi
Emmenopterys henryi
Hepatica henryi
Lilium henryi
Lonicera henryi
Parthenocissus henryana
Rubus henryi
Tilia henryana

hepta- seven- (see below)
heptalobus with seven lobes
   Acer palmatum var. *heptalobum*
heptaphyllus with seven leaves or
   leaflets
   Cardamine heptaphyllus
heracleifolius with leaves like
   hogweed (*Heracleum*)
   Clematis heracleifolia

CULTIVAR NAMES
Public Figures

NARCISSUS 'KING ALFRED'
LEUCANTHEMUM × *SUPERBUM*
   'BARBARA BUSH'
CLEMATIS 'PRINCE CHARLES'
FUCHSIA 'WINSTON CHURCHILL'
CLEMATIS 'PRINCESS DIANA'
COLCHICUM 'DISRAELI'
IRIS 'DUKE OF EDINBURGH'
CLEMATIS 'KING EDWARD VII'
CANNA 'GENERAL EISENHOWER'
ROSA 'THE QUEEN ELIZABETH'
PENSTEMON 'KING GEORGE V'
CLEMATIS 'W.E. GLADSTONE'
CLEMATIS 'JAN PAWEL' (POPE JOHN
   PAUL)
CLEMATIS 'LADY BIRD JOHNSON'
PAEONIA LACTIFLORA
   'LORD KITCHENER'
MAGNOLIA 'PRINCESS MARGARET'
HELIOTROPIUM ARBORESCENS
   'PRINCESS MARINA'
PRIMULA 'NAPOLEON'
ROSA 'CARDINAL DE RICHELIEU'
PAEONIA LACTIFLORA 'PRESIDENT
   FRANKLIN D. ROOSEVELT'
PRIMULA AURICULA 'ANWAR SADAT'
ROSA QUEEN MOTHER
   ('KORQUEMU')
LOBELIA 'QUEEN VICTORIA'
HELIOTROPIUM ARBORESCENS
   'THE SPEAKER'
CLEMATIS 'PRINCESS OF WALES'
JUNIPERUS HORIZONTALIS
   'PRINCE OF WALES'
SEMPERVIVUM 'DUKE OF WINDSOR'

herbaceus herbaceous, not woody; grass-green

hercegovinus from Hercegovina
   Helleborus multifidus subsp. *hercegovinus*

hesperus of the west; of evening

heter-, hetero- diverse, various

heteracanthus with differing spines

heteranthus with diverse flowers (i.e. different kinds – such as sterile and fertile – on one plant)
   Indigofera heterantha

heterocarpus with differing fruits

heterodontus with differing teeth
   Rhodiola heterodonta

heterodoxus differing from others in its genus
   Penstemon heterodoxus

heterolepis with differing scales

heteromorphus with differing forms

SIR HAROLD HILLIER (1905–1985)
The grandson of Edwin Hillier, who founded Hillier Nurseries in 1864, Sir Harold took over the business in 1944. An eminent plantsman, he created a world-class arboretum at his home, Jermyns House, Ampfield, and later in life travelled widely to study plants in the wild. Many plant names have Hillier associations. As well as the obvious ones (*Robinia* × *slavinii* 'Hillieri', *Viburnum* × *hillieri*, *Thuja plicata* 'Hillieri' etc.), *wintonensis* (from Winchester) usually indicates a Hillier plant (e.g. × *Halimiocistus wintonensis*). Several plants have the cultivar name 'Jermyns' (e.g. *Abutilon* × *suntense* 'Jermyns').

heterophyllus with diverse leaves
   (i.e. different forms on one plant)
Cirsium heterophyllum
Fagus sylvatica var. *heterophylla*
Osmanthus heterophyllus
Penstemon heterophyllus

hexa- six- (see below)
hexagonus six-angled
hexandrus with six stamens
   Podophyllum hexandrum
hibernicus Irish
   Hedera hibernica
hibernus of winter; Irish
hiemalis of winter, winter-flowering
hieraciifolius like *Hieracium*
   (hawkweed)
hillieri (see panel opposite)

COMMON DESCRIPTIVE TERMS
himalaicus, himalayensis from
   the Himalayas
Aster himalaicus
Eremurus himalaicus
Geranium himalayense
Incarvillea himalayensis

himanto- strap-shaped
hinnuleus fawn in colour
hippocastanus horse chestnut
   Aesculus hippocastanum
hippocrepis like a horseshoe
hircinus like a goat; smelling goaty

CULTIVAR NAMES
Strange but True
H

ALLIUM VINEALE 'HAIR'
PELARGONIUM 'HAPPY THOUGHT'
MALUS DOMESTICA (APPLE)
'HARRY MASTER'S JERSEY'
POLEMONIUM BOREALE
'HEAVENLY HABIT'
PANICUM VIRGATUM 'HEAVY METAL'
RANUNCULUS ACRIS 'HEDGEHOG'
IRIS 'HELLO DARKNESS'
RHODODENDRON ABERCONWAYI
'HIS LORDSHIP'
FUCHSIA 'HOBO'
GERANIUM PRATENSE 'HOCUS POCUS'
PHLOX STOLONIFERA 'HOME FIRES'
IRIS 'HONKY TONK BLUES'
RHODODENDRON 'HOPPY'
ORIGANUM 'HOT AND SPICY'
CAMPANULA PUNCTATA 'HOT LIPS'
HEMEROCALLIS 'HOUDINI'
EUPHORBIA CHARACIAS SUBSP.
CHARACIAS 'HUMPTY DUMPTY'
PINUS MUGO 'HUMPY'
PELARGONIUM 'HURDY-GURDY'

COMMON DESCRIPTIVE TERMS
hirsutus, hirtus hairy
Kniphofia hirsuta
Lotus hirsutus
Penstemon hirsutus
Polygonatum hirtum
Primula hirsuta
Rudbeckia hirta

CULTIVAR NAMES
Foreign Expressions
H

HEIDEBRAUT (*German*)
HEATHLAND BRIDE
(*MOLINIA CAERULEA* SUBSP. *CAERULEA*
'HEIDEBRAUT')

HERBSTFEUER (*German*)
AUTUMN FIRE
(*COTONEASTER SALICIFOLIUS*
'HERBSTFEUER')

HERBSTFREUDE (*German*)
AUTUMN JOY
(*SEDUM* 'HERBSTFREUDE')

HERBSTSCHNEE (*German*)
AUTUMN SNOW
(*ASTER NOVAE-ANGLIAE*
'HERBSTSCHNEE')

HERBSTSONNE (*German*)
AUTUMN SUN
(*RUDBECKIA LACINIATA*
'HERBSTSONNE')

HERBSTZAUBER (*German*)
AUTUMN MAGIC
(*PENNISETUM ALOPECUROIDES*
'HERBSTZAUBER')

HIGASAYAMA (*Japanese*)
PARASOL MOUNTAIN
(*ACER PALMATUM* 'HIGASAYAMA')

HIMMELBLAU (*German*)
SKY BLUE
(*PINUS STROBUS* 'HIMMELBLAU')

hispanicus Spanish
Angelica hispanica
Genista hispanica
Hyacinthoides hispanica
Platanus × *hispanica*
Quercus × *hispanica*
Sedum hispanicum

hispidus bristly
Carex hispida
Drosanthemum hispidum
Elymus hispidus
Heuchera hispida
Pterostyrax hispida
Robinia hispida

hollandicus from Holland
  Allium hollandicum
holo- entirely, completely, wholly
  (see below)
holocarpus with entire (i.e.
  undivided) fruits
holochrysus completely golden
  Phyllostachys bambusoides
  'Holochrysa'
hololeucus completely white
holophyllus with entire (i.e.
  undivided) leaves
  Abies holophylla
homo- like; of the same kind
  (see below)
homocarpus with fruit of only
  one kind

hirtellus rather hairy
hirtiflorus with hairy flowers
hirtipes with a hairy foot or stalk
  Rhododendron hirtipes

homolepis with uniform scales
   Abies homolepis
homomallus, homotropus all turned in the same direction
hondoensis from Hondo, Japan
hookeri, hookerianus (see panel above)
horizontalis horizontal, spreading horizontally
   Cotoneaster horizontalis
hormophorus with a necklace
horricomis bristly, shaggy
horridulus standing up, projecting
   Meconopsis horridula
horridus standing on end; prickly
   Eryngium horridum
hortensis of gardens, raised in a garden
   Atriplex hortensis
hortorum of gardens
hortulanorum of gardeners

hugonis named after Father Hugh (1851–1928), an Irish missionary who spent many years in China
   Rosa xanthina f. *hugonis*
humectus moist, damp
humifusus spread on the ground
   Sedum humifusum

COMMON DESCRIPTIVE TERMS

humilis dwarf, low-growing
Chamaerops humilis
Jasminum humile
Penstemon humilis
Phyllostachys humilis
Pleioblastus humilis
Polygonatum humile

---

humulifolius with leaves like *Humulus* (hop)
hungaricus Hungarian
   Acanthus hungaricus

Hackberry *Celtis*
Hair grass *Deschampsia*
Handkerchief tree *Davidia involucrata*
Hard fern *Blechnum spicant*
Hardheads *Centaurea nigra*
Hardy plumbago *Ceratostigma*
Hare's tail *Lagurus ovatus*
Harebell *Campanula rotundifolia*
Harry Lauder's walking stick *Corylus avellana* 'Contorta'
Hart's tongue fern *Asplenium scolopendrium*
Hattie's pincushion *Astrantia*
Hawksbeard *Crepis*
Hawkweed *Hieracium*; *Pilosella*
Hawthorn *Crataegus*
Hazel *Corylus avellana*
Heartsease *Viola tricolor*
Heath *Erica*; *Daboecia*
Heather *Calluna*
Hedgehog holly *Ilex aquifolium* 'Ferox'
Heliotrope *Heliotropium arborescens*
Hellebore *Helleborus*
Hemlock *Tsuga*; *Conium maculatum*
Hemp agrimony *Eupatorium cannabinum*
Hen-and-chickens houseleek *Jovibarba sobolifera*
Herb Robert *Geranium robertianum*
(Herbs, see page 47)
Hiba *Thujopsis dolabrata*
Hickory *Carya*
Highclere holly *Ilex* × *altaclerensis*
Himalayan birch *Betula utilis*
Himalayan blue poppy *Meconopsis betonicifolia*
Himalayan pine *Pinus wallichiana*
Hinoki cypress *Chamaecyparis obtusa*

Hogweed *Heracleum sphondylium*
Holly *Ilex*
Hollyhock *Alcea rosea*
Holm oak *Quercus ilex*
Honesty *Lunaria annua*
Honey bush *Melianthus major*
Honey locust *Gleditsia triacanthos*
Honeysuckle *Lonicera*
Hop *Humulus lupulus*
Hop hornbeam *Ostrya carpinifolia*
Hornbeam *Carpinus betulus*
Horned-poppy *Glaucium*
Horse chestnut *Aesculus hippocastanum*
Horsetail *Equisetum*
Hot water plant *Achimenes*
Hottentot fig *Carpobrotus edulis*
Hound's tongue *Cynoglossum*
Houseleek *Sempervivum*
Huckleberry *Gaylussacia*

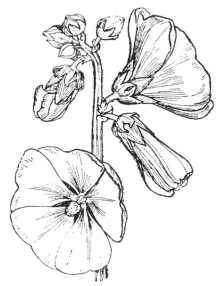

HOLLYHOCK

hupehensis from Hubei province
in China
Anemone hupehensis
Malus hupehensis
Sorbus hupehensis

hyacinthinus deep purplish-blue, like
Hyacinthus
hyacinthoides like *Hyacinthus*
hyalinus colourless, transparent
hybernus Irish
Euphorbia hyberna

hybridus hybrid; of mixed parentage
Anemone × *hybrida*
Arctotis × *hybrida*
Cistus × *hybridus*
Helleborus × *hybridus*
Polygonatum × *hybridum*
Rheum × *hybridum* (rhubarb)

hydrangeoides like *Hydrangea*
hydro- water-
hydrophilus water-loving
hyemalis of winter
Eranthis hyemalis
hygro- damp
hylaeus of woods
hymen- membrane
hymenodes like a membrane
hymenorrhizus with membranous
roots

GENUS NAMES

H

Hakonechloa named for Mount
Hakone, in Japan
Hebe named for the Greek goddess
of youth
Hedychium from Greek *hedys*, sweet,
and *chion*, snow
Helianthus from Greek *helios*, sun,
and *anthos*, flower
Hemerocallis from Greek *hemera*, day,
and *kallos*, beauty
Hepatica from Greek *hepar*, liver
Hesperis from Greek *hespera*, evening
Heuchera named for Johann von
Heucher (1677–1747), a German
professor
Hosta named for Nicolaus Host
(1761–1834), Austrian doctor

hymenosepalus with membranous
sepals
hyper- beyond, above
hyperboreus from the far north
hypericoides like *Hypericum*
hypnoides like moss
hypo- beneath (see below)
hypogaeus growing underground
hypoglaucus glaucous beneath
hypoleucus white beneath
hyrcanus from Hyrcania, an old
region near the Caspian Sea
hyssopifolius with leaves like the herb
Hyssopus officinalis (hyssop)
hystrix porcupine; spiny
Picea abies 'Hystrix'

I

ibericus from the Iberian peninsula
  Geranium ibericum
icterinus jaundice-yellow
  Salvia officinalis 'Icterina'
idahoensis from Idaho, USA
  Sisyrinchium idahoense

CULTIVAR NAMES
Personal Names
H, I

CLEMATIS VITICELLA 'HANNA'
HAMAMELIS × *INTERMEDIA* 'HARRY'
LAVANDULA 'HAZEL'
GERANIUM CINEREUM 'HEATHER'
ACHILLEA 'HEIDI'
PRIMULA AURICULA 'HELEN'
EREMURUS 'HELENA'
HIBISCUS SYRIACUS 'HELENE'
ITEA VIRGINICA 'HENRY'S GARNET'
VACCINIUM CORYMBOSUM
  'HERBERT'
PRUNUS DOMESTICA (PLUM)
  'HERMAN'
FUCHSIA 'HOLLY'S BEAUTY'
SEDUM TELEPHIUM 'HESTER'
IRIS 'IDA'
PELARGONIUM 'DARK RED IRENE'
CAMPANULA CARPATICA 'ISABEL'
ERICA CARNEA 'ISABELL'
ROSA ISABELLA ('POULISAB')
GERANIUM 'IVAN'

---

ida-maia named for Ida May Burke,
  a 19th-century plant hunter in
  California
igneus fiery red
  Cuphea ignea
ikariae from the Greek island, Ikaria
  Galanthus ikariae
ilicifolius with leaves like *Ilex* (holly)
  Itea ilicifolia
illinoinensis from Illinois, USA
  Carya illinoinensis (pecan)
illustris brilliant, clear, lit up
  Amsonia illustris
illyricus from Illyria (the Adriatic
  coast of Croatia and Dalmatia)
  Gladiolus illyricus
imbricatus overlapping regularly,
  like tiles
  Fabiana imbricata
impeditus hindered, obstructed
  Rhododendron impeditum

COMMON DESCRIPTIVE TERMS
imperialis imperial, showy
Alnus glutinosa 'Imperialis'
Dahlia imperialis
Fritillaria imperialis

---

implexus entangled, interwoven
  Lonicera implexa
impressus impressed, sunken
  Ceanothus impressus
inapertus closed
  Agapanthus inapertus

WHAT'S IN A NAME?

SHAPE

acerosus needle-shaped
acetabulosus cup-shaped, concave
acicularis like a pin or needle
aciformis needle-shaped
acuminatus tapering to a point
alveolatus hollowed out, channelled
annularis ring-shaped
apiculatus with a short pointed end
applanatus flattened
arcuatus bent or curved like a bow
attenuatus tapering to a point
bucinatus like a curved horn
calathinus cup-shaped; like a basket
calceiformis, calceolatus shaped like a
   little shoe or slipper
campanulatus bell-shaped
canaliculatus with a channel, like a pipe
cassideus shaped like a helmet
catilliformis shaped like a saucer
caudiformis shaped like a tail
clathratus like a lattice or trellis
clavatus, claviformis club-shaped
conicus cone-shaped
convolutus rolled up lengthways
cordatus, cordiformis heart-shaped
corniculatus with little horns
cornutus horned
cristatus crested
cruciatus, cruciformis cross-shaped
cucullatus hooded
curvulus slightly curved
cuspidatus with a sharp point
cyathiformis cup-shaped
cylindricus, cylindraceus cylindrical
deltoides triangular
depressus flattened
ellipticus elliptic
ensatus, ensiformis sword-shaped

falcatus falcate, curved like a sickle
flabellatus fan-shaped
flexuosus, fractiflexus zigzag
fornicatus arched
fungiformis mushroom-shaped
fusiformis shaped like a spindle
gibbosus hunchbacked
lanceolatus lanceolate, spear-shaped
lunulatus like a crescent moon
mucronatus pointed
nidiformis nest-shaped
nummularius like a coin
obconicus like an inverted cone
orbiculatus, orbicularis disc-shaped
ovalis oval, broadly elliptic
ovatus ovate, egg-shaped
peltatus like a pelta, a small, half-moon-
   shaped shield
planus flat
platy- broad, flat
poculiformis cup-shaped
prismaticus prism-shaped
pulvinaris, pulvinatus like a cushion
pungens ending in a hard sharp point
pyramidalis pyramid-shaped
reniformis kidney-shaped
rhombeus, rhombicus, rhombiformis
   rhomboidal, diamond-shaped
rotatus shaped like a wheel
saccatus like a bag; with sacs
sagittalis, sagittatus arrow-shaped
sphaerocephalus with globose heads
spicatus spicate, bearing a spike
stellatus stellate, star-shaped
strepto- twisted
tabularis, tabuliformis flat like a board
uncinatus hooked, with a hooked end
undulatus, undatus wavy

WHAT'S IN A NAME?

COLOUR: BROWN

badius reddish brown
brunneus dark, dull brown
castaneus chestnut
cervinus fawn
cinnamomeus cinnamon brown
cupreus coppery
ferrugineus; ferruginosus rusty; light
  reddish brown
fuligineus; fuliginosus dark brown,
  sooty
fuscus dark; blackish brown
glandulaceus tawny brown
hepaticus liver-coloured
luridus dirty brown
testaceus terracotta-coloured
umbrinus umber

COMMON DESCRIPTIVE TERMS

incanus grey, hoary
Alnus incana
Crepis incana
Geranium incanum
Marrubium incanum
Matthiola incana
Scutellaria incana

incarnatus flesh-coloured
  Passiflora incarnata
incisus deeply cut
  Prunus incisa
incompletus incomplete
  Paris incompleta
incomptus unadorned
  Pennisetum incomptum

incurvus curved inwards
  Campanula incurva

COMMON DESCRIPTIVE TERMS

indicus Indian (often used more
  generally to mean from the
  Far East)
Aesculus indica
Canna indica
Duchesnea indica
Ipomoea indica
Plumbago indica

indivisus undivided
  Cordyline indivisa
inermis unarmed (i.e. without
  prickles)
  Bromus inermis 'Skinner's Gold'
inflatus inflated, blown up, swollen
  Cistus inflatus
inflexus bent inwards
  Juncus inflexus
innominatus unnamed
  Iris innominata
inodorus scentless
  Hypericum × inodorum

INCISUS

insignis distinguished
  Rhododendron insigne
insititius grafted
insuetus unusual
insularis of islands, insular
  Fraxinus insularis var. *henryana*
intaminatus undefiled, pure
  Cyclamen intaminatum
integer entire, undivided
  Salix integra
integerrimus completely entire
  Diascia integerrima

INTEGER

CULTIVAR NAMES
Special Occasions

FUCHSIA 'BICENTENNIAL'
FUCHSIA 'BRITISH JUBILEE'
IRIS 'CARNIVAL TIME'
FUCHSIA 'CELEBRATION'
HOSTA 'CELEBRATION'
CEANOTHUS 'CENTENNIAL'
HEMEROCALLIS 'CHILDREN'S FESTIVAL'
IRIS SIBIRICA 'CORONATION ANTHEM'
RHODODENDRON 'CORONATION DAY'
ACHILLEA 'CORONATION GOLD'
DIANTHUS 'CORONATION RUBY'
ROSA 'DIAMOND JUBILEE'
FUCHSIA 'EASTER BONNET'
NARCISSUS 'EASTER BONNET'
PRIMULA 'EASTER BONNET'
LIQUIDAMBAR STYRACIFLUA 'FESTIVAL'
ERYSIMUM 'GOLDEN JUBILEE'
CAMELLIA JAPONICA 'GRAND PRIX'
HELIANTHEMUM 'JUBILEE'

ROSA 'JUBILEE CELEBRATION'
GERANIUM SANGUINEUM 'JUBILEE PINK'
RHODODENDRON 'MARDI GRAS'
RHODODENDRON 'MAY DAY'
FUCHSIA 'MILLENNIUM'
RHODODENDRON 'MOTHER'S DAY'
PHYSOSTEGIA VIRGINIANA 'OLYMPIC GOLD'
PAPAVER ORIENTALE 'ROYAL WEDDING'
NARCISSUS 'SAINT PATRICK'S DAY'
ROSA 'SCOTTISH CELEBRATION'
OZOTHAMNUS ROSMARINIFOLIUS 'SILVER JUBILEE'
ROSA 'SILVER JUBILEE'
CLEMATIS 'SPECIAL OCCASION'
RHODODENDRON 'WASHINGTON STATE CENTENNIAL'
GYPSOPHILA FASTIGIATA 'WHITE FESTIVAL'

integrifolius with undivided leaves
Calceolaria integrifolia
Clematis integrifolia
Cotoneaster integrifolius
Hydrangea integrifolia
Meconopsis integrifolia

interjectus placed between,
  intermediate
  Polypodium interjectum

GENUS NAMES
I, J, K

Impatiens in Latin, impatient
Incarvillea named for Pierre
  d'Incarville (1706–1757), French
  missionary and botanist
Iris Roman goddess of the rainbow
Jovibarba from medieval Latin
  meaning beard of Jupiter
Kalmia named for Pehr Kalm (1715–
  1779), a Finn who travelled in North
  America and studied with Linnaeus
Kerria named for William Kerr
  (d. 1814), Scottish horticulturist at
  Kew and plant collector in China
Knautia named for Christoph Knaut
  (1638–1694), German physician
  and botanist
Kniphofia named for Johann Kniphof
  (1704–1763), German professor
  and botanical author
Kolkwitzia named for Richard
  Kolkwitz (1873–1956), German
  botanist

intermedius intermediate (in form,
  colour etc.; usually applied to a
  hybrid that is intermediate between
  its parents)
Eucryphia × *intermedia*
Forsythia × *intermedia*
Hamamelis × *intermedia*
Lavandula × *intermedia*

interruptus scattered, not continuous
  Salvia interrupta
intricatus tangled
  Clematis intricata

involucratus having an involucre
  (a circle of bracts around the
  flowers)
Astrantia major subsp. *involucrata*
Cyperus involucratus
Davidia involucrata
Fritillaria involucrata
Hydrangea involucrata
Lonicera involucrata
Salvia involucrata

iridescens iridescent
  Phyllostachys iridescens
iridiflorus with flowers like *Iris*
  Canna iridiflora
irroratus sprinkled with dew; finely
  spotted
  Salix irrorata

INDIANERSOMMER (*German*)
INDIAN SUMMER
(*HELENIUM* 'INDIANERSOMMER')

JULIGLUT (*German*) JULY GLOW
(*PHLOX PANICULATA* 'JULIGLUT')

KAGIRI-NISHIKI (*Japanese*)
GLORIOUS COLOUR
(*ACER PALMATUM* 'KAGIRI-NISHIKI')

KIKU-SHIDARE-ZAKURA (*Japanese*)
CHRYSANTHEMUM WEEPING CHERRY
(*PRUNUS* 'KIKU-SHIDARE-ZAKURA')

KISSEN (*German*) CUSHION
(*PINUS MUGO* 'KISSEN')

KLEINE LIEBLING (*German*)
LITTLE DARLING
(*PELARGONIUM* 'KLEINE LIEBLING')

KLEINE SILBERSPINNE (*German*)
LITTLE SILVER SPIDER
(*MISCANTHUS SINENSIS* 'KLEINE
SILBERSPINNE')

KLEINE TÄNZERIN (*German*)
LITTLE DANCER
(*PAPAVER ORIENTALE* 'KLEINE
TÄNZERIN')

KLEINER FUCHS (*German*)
LITTLE FOX
(*HELENIUM* 'KLEINER FUCHS')

KOJO-NO-MAI (*Japanese*)
DANCE IN THE ANCIENT CASTLE
(*PRUNUS INCISA* 'KOJO-NO-MAI')

KÖNIGIN (*German*) QUEEN
(E.G. *ECHINACEA PURPUREA*
'AUGUSTKÖNIGIN')

KÖNIGSKIND (*German*) KING'S CHILD
(*CLEMATIS* 'KÖNIGSKIND')

KONINGIN (*Dutch*) QUEEN
(*TRITELEIA LAXA* 'KONINGIN FABIOLA')

KRASAVITSA MOSKVY (*Russian*)
BEAUTY OF MOSCOW
(*SYRINGA VULGARIS* 'KRASAVITSA
MOSKVY')

KUMO-NO-OBI (*Japanese*)
BEAR'S BELT
(*IRIS ENSATA* 'KUMO-NO-OBI')

KUPFERTEPPICH (*German*)
COPPER CARPET
(*ACAENA MICROPHYLLA*
'KUPFERTEPPICH')

---

isabellinus tawny yellow
  Eremurus × *isabellinus*
isophyllus with equal leaves
  Campanula isophylla
italicus (see panel, right)
itinerans travelling
ixioides like *Ixia*

COMMON DESCRIPTIVE TERMS
italicus Italian
Arum italicum
Gladiolus italicus
Helichrysum italicum
Phlomis italica
Populus nigra 'Italica'

J

jackmanii named for George Jackman (1801–1869) and his son George (1837–1887), pioneering British clematis breeders at their nursery in Woking, Surrey
Clematis 'Jackmanii'

COMMON DESCRIPTIVE TERMS
jacquemontii, jacquemontianus named for the French naturalist Victor Jacquemont (1801–1832)
Arisaema jacquemontii
Betula utilis var. *jacquemontii*
Euphorbia jacquemontii
Parrotiopsis jacquemontiana

CULTIVAR NAMES
World of Sport
NARCISSUS 'ARKLE'
LUPINUS ARBOREUS 'ASTON VILLA'
PRIMULA AURICULA 'DAVID BECKHAM'
FUCHSIA 'SIR MATT BUSBY'
ROSA 'BOBBY CHARLTON'
FUCHSIA 'KENNY DALGLISH'
PRIMULA AURICULA 'GAZZA'
ROSA 'JIMMY GREAVES'
DAHLIA 'SIR ALF RAMSEY'
PRIMULA AURICULA 'RED RUM'
CROCUS CHRYSANTHUS 'ARD SCHENK'
FUCHSIA 'TORVILL AND DEAN'

---

jalapa from Jalapa in Mexico
Mirabilis jalapa

COMMON DESCRIPTIVE TERMS
japonicus Japanese
Acer japonicum
Aucuba japonica
Camellia japonica
Chaenomeles japonica
Cryptomeria japonica
Euonymus japonicus
Fatsia japonica
Kerria japonica
Pieris japonica
Primula japonica
Skimmia japonica
Spiraea japonica

---

jasminoides like *Jasminum* (jasmine)
Trachelospermum jasminoides
javanicus from Java
Oenanthe javanica
jeffreyi named for the Scottish plant collector John Jeffrey (1826–1854)
Pinus jeffreyi
jonquilla like *Juncus* (rush)
Narcissus jonquilla
jouinianus named for Emile Jouin, nursery manager at Simon-Louis Frères in Metz, France, where the plant was raised
Clematis × *jouiniana*
jubatus with a mane or crest
Hordeum jubatum

AGAPANTHUS 'JACK'S BLUE'
VERBASCUM 'JACKIE'
ERYSIMUM 'JACOB'S JACKET'
DIASCIA 'JACQUELINE'S JOY'
CLEMATIS 'JACQUI'
HEDERA HELIX 'JAKE'
PLECTRANTHUS FRUTICOSUS 'JAMES'
MAGNOLIA 'JANE'
HOSTA 'JANET'
FUCHSIA 'JANICE ANN'
ERICA CARNEA 'JEAN'
ERYTHRONIUM 'JEANNINE'
SCHIZOSTYLIS COCCINEA 'JENNIFER'
NARCISSUS 'JENNY'
FUCHSIA 'JESS'
CROCOSMIA × *CROCOSMIIFLORA* 'JESSIE'
PAPAVER ORIENTALE 'BIG JIM'
IRIS 'BROADLEIGH JOAN'
ERYTHRONIUM 'JOANNA'

EREMURUS 'JOANNE'
ROSA 'JOCELYN'
DIANTHUS 'LITTLE JOCK'
CLEMATIS × *CARTMANNII* 'JOE'
PRIMULA AURICULA 'JOEL'
PRIMULA 'JOHANNA'
IRIS 'JOHN'
PRIMULA 'JO-JO'
CLEMATIS JOSEPHINE ('EVIJOHILL')
GERANIUM 'JOY'
IRIS 'JOYCE'
MAGNOLIA 'JUDY'
ROSA 'JULIA'S ROSE'
VIOLA 'JULIAN'
GERANIUM NODOSUM 'JULIE'S VELVET'
ROSA SWEET JULIET ('AUSLEAP')
COTONEASTER × *SUECICUS* 'JULIETTE'
FUCHSIA 'JUSTIN'S PRIDE'

---

jucundus pleasant, delightful
  Osteospermum jucundum
juddii named for William Judd
  (1888–1946), propagator at the
  Arnold Arboretum in the USA
  Viburnum × *juddii*
junceus like *Juncus* (rush)
  Spartium junceum
juncifolius with rushlike leaves
juniperifolius with leaves like
  Juniperus (juniper)
  Armeria juniperifolia

juniperinus like juniper
  Grevillea juniperina

CAMELLIA JAPONICA

K

kaempferi named for a German traveller and botanical author, Engelbert Kaempfer (1651–1716)
Larix kaempferi
kaki persimmon (Japanese)
Diospyros kaki
kalmiiflorus with flowers like *Kalmia*
Deutzia kalmiiflora

CULTIVAR NAMES
Personal Names
K

HELIANTHEMUM 'KAREN'S SILVER'
PAPAVER ORIENTALE 'KARINE'
PHLOX PANICULATA 'KATARINA'
GERANIUM 'KATE'
LAVANDULA ANGUSTIFOLIA 'MISS KATHERINE'
ROSA 'KATHLEEN'
CAMPANULA CARPATICA 'KATHY'
CAMELLIA JAPONICA 'KATIE'
FUCHSIA 'KATRINA'
DAHLIA 'KEITH'S CHOICE'
CAMELLIA JAPONICA 'KENNY'
CALLUNA VULGARIS 'KERSTIN'
IRIS 'KEVIN'S THEME'
ECHINACEA 'KIM'S KNEE HIGH'
CAMELLIA JAPONICA 'KIMBERLEY'
LIQUIDAMBAR STYRACIFLUA 'KIRSTEN'
NARCISSUS 'KITTY'
ASTER 'KYLIE'

---

kamtschaticus from Kamchatka
Sedum kamtschaticum
kansuensis from Gansu province in north-western China
Malus kansuensis
karataviensis, karatavicus from the Karatau Mountains of Kazakhstan
Allium karataviense
Thymus karatavicus
karvinskianus named for Wilhelm Karwinsky von Karwin, a 19th-century German explorer
Erigeron karvinskianus
kashmirianus, kashmiriensis from Kashmir
Iris kashmiriana
kaufmannianus named for General von Kaufmann, a governor in Kazakhstan
Tulipa kaufmanniana
kermesinus carmine, purplish red
Clematis 'Kermesina'
kerneri, kernerianus named for Johann von Kerner (1755–1830), a German botanist; or for Anton Kerner von Marilaun (1831–1898), an Austrian botanist and author

---

COMMON DESCRIPTIVE TERMS
kewensis named for the Royal Botanic Gardens at Kew, London
Buddleja colvilei 'Kewensis'
Cytisus × *kewensis*
Sorbus × *kewensis*

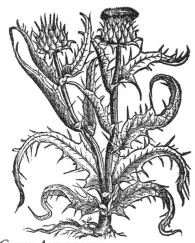

GLOBE ARTICHOKE

Artichoke, globe *Cynara scolymus*
  Scolymus Group
Artichoke, Jerusalem *Helianthus
  tuberosus*
Asparagus *Asparagus officinalis*
Aubergine *Solanum melongena*
Beans, broad *Vicia faba*
Beans, French *Phaseolus vulgaris*
Beans, runner *Phaseolus coccineus*
Beetroot *Beta vulgaris*
Broccoli *Brassica oleracea* Italica Group
Brussels sprouts *Brassica oleracea*
  Gemmifera Group
Cabbage *Brassica oleracea* Capitata
  Group
Carrot *Daucus carota*
Cauliflower *Brassica oleracea* Botrytis
  Group
Celeriac *Apium graveolens* var. *rapaceum*
Celery *Apium graveolens* var. *dulce*
Chard *Beta vulgaris*
Chicory *Cichorium intybus*

Chilli pepper *Capsicum annuum*
Courgette *Cucurbita pepo*
Cucumber *Cucumis sativus*
Endive *Cichorium endivia*
Fennel (Florence) *Foeniculum vulgare*
  var. *azoricum*
Garlic *Allium sativum*
Horseradish *Armoracia rusticana*
Kale *Brassica oleracea* Acephala Group
Kohl rabi *Brassica oleracea* Gongylodes
  Group
Lamb's lettuce *Valerianella locusta*
Leek *Allium ampeloprasum porrum*
Lettuce *Lactuca sativa*
Marrow *Cucurbita pepo*
Okra *Abelmoschus esculentus*
Onion *Allium cepa*
Pak choi *Brassica rapa* Chinensis Group
Parsnip *Pastinacea sativa*
Peas *Pisum sativum*
Pepper *Capsicum annum*
Potato *Solanum tuberosum*
Radish *Raphanus sativus*
Rocket *Eruca sativa*
Salsify *Tragopogon porrifolium*
Scorzonera *Scorzonera hispanica*
Shallot *Allium cepa* Aggregatum Group
Spinach *Spinacia oleracea*
Squash *Cucurbita pepo; Cucurbita
  maxima; Cucurbita moschata*
Swede *Brassica napus* Napobrassica
  Group
Sweet potato *Ipomoea batatas*
Sweetcorn *Zea mays*
Tomato *Lycopersicon esculentum*
Turnip *Brassica napus* Rapifera Group
Watercress *Nasturtium officinalis*
Welsh onion *Allium fistulosum*

Ice plant *Sedum spectabile*
Iceland poppy *Papaver nudicaule*
Incense cedar *Calocedrus decurrens*
Indian bean *Lablab purpureus*
Indian bean tree *Catalpa bignonoides*
Indian horse chestnut *Aesculus indica*
Indian paint-brush *Castilleja*
Indian shot *Canna*
Irish yew *Taxus baccata* 'Fastigiata'
Ironbark *Eucalyptus*
Ironwood *Ostrya virginiana*
Italian alder *Alnus cordata*
Italian cypress *Cupressus sempervirens*
Ivy *Hedera*
Jack-go-to-bed-at-noon *Tragopogon pratensis*
Jacob's ladder *Polemonium caeruleum*
Jacob's rod *Asphodeline lutea*
Jade plant *Crassula ovata*
Japanese anemone *Anemone* × *hybrida*; *Anemone hupehensis*
Japanese angelica tree *Aralia elata*
Japanese cedar *Cryptomeria japonica*
Japanese cherries *Prunus* (many named forms)
Japanese crab *Malus floribunda*
Japanese maples *Acer palmatum*
Japanese quince, Japonica *Chaenomeles*
Japanese snowbell *Styrax japonica*
Japanese strawberry tree *Cornus kousa*
Jasmine, Jessamine *Jasminum*
Jerusalem sage *Phlomis fruticosa*
Joe Pye weed *Eupatorium purpureum*
Jonquil *Narcissus jonquilla*
Joshua tree *Yucca brevifolia*
Judas tree *Cercis siliquastrum*
June berry *Amelanchier*
Juniper *Juniperus*

Kaffir lily *Schizostylis*
Kansas gayfeather *Liatris*
Kashmir cypress *Cupressus cashmeriana*
Katsura tree *Cercidiphyllum japonicum*
Keaki *Zelkova serrata*
Kentucky coffee tree *Gymnocladus dioica*
Kerosene bush *Ozothamnus ledifolius*
Kilmarnock willow *Salix caprea* 'Kilmarnock'
Kingcup *Caltha palustris*
Knapweed *Centaurea*
Knotweed *Persicaria*
Korean fir *Abies koreana*
Kowhai *Sophora microphylla*
Kumquat *Fortunella*

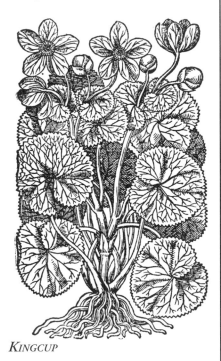

KINGCUP

kishtvariensis named for a valley in Kashmir
   Geranium kishtvariense
kitaibelii named for a Hungarian botanist, Paul Kitaibel (1757–1817)
   Cardamine kitaibelii
kiusianus from Kyushu, Japan
   Veronica kiusiana
koehneanus named for Bernhard Koehne (1848–1918), a German botanist
   Sorbus koehneana
kokanicus from Khokand, Turkestan
   Morina kokanica
komarovii named for Vladimir Komarov (1869–1945), a Russian botanist and explorer in Siberia and Korea
   Syringa komarovii

COMMON DESCRIPTIVE TERMS
koreanus, koraiensis from Korea
Abies koreana
Clematis koreana
Geranium koraiense
Geranium koreanum
Thuja koraiensis

---

kotschyanus, kotschyi named for Theodore Kotschy (1813–1866), an Austrian botanist and plant collector
   Erysimum kotschyanum

CULTIVAR NAMES
Strange but True
I, J, K
HEMEROCALLIS 'INDIAN PAINTBRUSH'
TIARELLA 'INKBLOT'
CROCOSMIA × *CROCOSMIIFLORA* 'JACKANAPES'
ROSA THE INGENIOUS MR FAIRCHILD ('AUSTIJUS')
SOLENOSTEMON 'INKY FINGERS'
SILENE 'JACK FLASH'
BRUNNERA MACROPHYLLA 'JACK FROST'
PHORMIUM 'JACK SPRATT'
IRIS 'JAZZAMATAZZ'
TIARELLA 'JEEPERS CREEPERS'
HEMEROCALLIS 'JENNY WREN'
GERANIUM 'JOLLY BEE'
SEMPERVIVUM 'JOLLY GREEN GIANT'
PENSTEMON 'THE JUGGLER'
NARCISSUS 'JUMBLIE'
IRIS 'JURASSIC PARK'
PHLOX 'KELLY'S EYE'
IRIS 'KILT LILT'
SEMPERVIVUM 'KIP'
FUCHSIA 'KISS 'N' TELL'
IRIS 'KNICK KNACK'
PAEONIA LACTIFLORA 'KRINKLED WHITE'

kouytchensis from Guizhou province in China
   Hypericum kouytchense
kurdicus from Kurdistan
   Tulipa kurdica

L

laceratus lacerated, torn
   Asplenium scolopendrium Laceratum
   Group

---

laciniatus, laciniosus cut into
   narrow divisions, jagged
Alnus glutinosa 'Laciniata'
Betula pendula 'Laciniata'
Sambucus nigra f. *laciniata*
Rudbeckia laciniata

---

lacteus milky
   Cotoneaster lacteus
lactiflorus with milky-coloured
   flowers
   Campanula lactiflora
lacustris of lakes
   Iris lacustris
ladakhianus from Ladakh in
   northern India
   Clematis ladakhiana

LACINIATUS

ladanifer bearing a medicinal resin
   called ladanum
   Cistus ladanifer
laetiflorus with bright flowers
   Helianthus × *laetiflorus*
laetus bright
   Pseudopanax laetus

---

laevigatus smooth
Crataegus laevigata
Crocus laevigatus
Hosta laevigata
Origanum laevigatum

laevis soft
Amelanchier laevis
Aster laevis
Escallonia laevis
Moluccella laevis

---

lagocephalus with a head like a hare
   Artemisia lagocephalus
lamarckii named for the French
   naturalist Chevalier Jean-Baptiste
   de Monet Lamarck (1744–1829)
   Amelanchier lamarckii

---

lanatus woolly
Digitalis lanata
Lavandula lanata
Phlomis lanata
Salix lanata

---

lancasteri named for the eminent British plantsman, broadcaster, lecturer and botanical author Roy Lancaster

Hypericum lancasteri

COMMON DESCRIPTIVE TERMS

lanceolatus lanceolate, spear-shaped

Azara lanceolata

Codonopsis lanceolata

Coreopsis lanceolata

Drimys lanceolata

Hoya lanceolata

Ornithogalum lanceolatum

Plantago lanceolata

---

lancifolius lance-leaved

  Hosta lancifolia

langsdorffii named for Georg Langsdorf (1774–1852), a German physician and botanist

  Nicotiana langsdorffii

laniger woolly

  Leptospermum lanigerum

lanuginosus woolly

  Androsace lanuginosa

lappa a bur

  Arctium lappa

laricifolius with leaves like larch

  Minuartia laricifolia

lasiandrus with woolly stamens

  Oxalis lasiandrus

lasianthus with woolly flowers

  Halimium lasianthum

CULTIVAR NAMES
Foreign Expressions

L

LA SÉDUISANTE (*French*) TEMPTRESS (*HEBE* 'LA SÉDUISANTE')

LACHSKÖNIGIN (*German*) SALMON QUEEN (*ASTILBE* 'LACHSKÖNIGIN')

LACHSSCHÖNHEIT (*German*) SALMON BEAUTY (*ACHILLEA* 'LACHSSCHÖNHEIT')

LANDHOCHZEIT (*German*) COUNTRY WEDDING (*PHLOX PANICULATA* 'LANDHOCHZEIT')

LANG TIDLIG ... (*Danish*) EARLY LONG ... (*CORYLUS* 'LANG TIDLIG ZELLER')

LAVENDELWOLKE (*German*) LAVENDER CLOUD (*PHLOX PANICULATA* 'LAVENDELWOLKE')

LE PHARE (*French*) LIGHTHOUSE (*PENSTEMON* 'LE PHARE')

LEUCHTFEUER (*German*) FIRELIGHT (*PAPAVER ORIENTALE* 'LEUCHTFEUER')

LEUCHTFUNK (*German*) (USUALLY TRANSLATED AS) FLASHING LIGHT (*DIANTHUS DELTOIDES* 'LEUCHTFUNK')

LEUCHTKÄFER (*German*) FIREFLY (*HEUCHERA* 'LEUCHTKÄFER')

CULTIVAR NAMES
Personal Names
L

CLEMATIS 'LAURA'

PAPAVER ORIENTALE 'LAUREN'S LILAC'

GALANTHUS 'LAVINIA'

FUCHSIA 'LENA'

HIBISCUS SYRIACUS 'LENNY'

LONICERA SEMPERVIRENS 'LEO'

FUCHSIA 'LEONORA'

CAMPANULA PORTENSCHLAGIANA 'LIESELOTTE'

GAULTHERIA MUCRONATA 'LILIAN'

RHODODENDRON 'LINDA'

PRIMULA AURICULA 'LISA'S SMILE'

FUCHSIA 'LIZ'

FUCHSIA LIZA ('GOETZLIZA')

PHLOX PANICULATA 'LIZZY'

RHODODENDRON 'LORNA'

LAVANDULA ANGUSTIFOLIA LITTLE LOTTIE ('CLARMO')

HOSTA 'LOUISA'

HEBE 'LOUISE'

CLEMATIS 'LUCIE'

PELARGONIUM 'LUCILLA'

ALSTROEMERIA 'LUCINDA'

PELARGONIUM 'LUCY'

PENSTEMON 'LYNETTE'

COMMON DESCRIPTIVE TERMS

lasiocarpus with woolly fruits

Abies lasiocarpa

Campanula lasiocarpa

Musa lasiocarpa

Populus lasiocarpa

---

lateriflorus with flowers at the side
  Aster lateriflorus
lateritius brick red
  Malvastrum lateritium

COMMON DESCRIPTIVE TERMS

latifolius broad-leaved

Campanula latifolia

Kalmia latifolia

Lavandula latifolia

Lathyrus latifolius

Limonium latifolium

Muscari latifolium

Phillyrea latifolia

---

latilobus with wide lobes
  Campanula latiloba
latisepalus with broad sepals
  Epimedium latisepalum
laureola a laurel crown or garland
  Daphne laureola
laurifolius with leaves like *Laurus* (bay)
  Cistus laurifolius
laurocerasus cherry laurel – from *laurus* (laurel) and *cerasus* (cherry)
  Prunus laurocerasus
lavalleei named for Pierre Lavalle (1836–1884), a French botanist
  Crataegus × *lavalleei*
lavandulifolius with leaves like *Lavandula* (lavender)
  Salvia lavandulifolia

lawsonianus named for the Scottish botanical author and nurseryman Charles Lawson (1794–1873)
  Chamaecyparis lawsoniana

COMMON DESCRIPTIVE TERMS

laxiflorus loose-flowered
  Campanula laxiflora
  Lathyrus laxiflorus
  Lobelia laxiflora

---

laxifolius loose-leaved
  Athrotaxis laxifolia
laxus loose, open
  Triteleia laxa 'Koningin Fabiola'
lazicus from Lazistan, on the Black Sea
  Iris lazica
ledifolius with leaves like the aromatic evergreen shrub *Ledum*
  Ozothamnus ledifolius

WHAT'S IN A NAME?
ANIMAL CONNECTIONS

anguilliformis eel-shaped
anguinus snakelike
anserinus relating to geese
apianus of bees
aquilinus eagle-like
arachnoides like a spider or its web
aucuparius bird-catching
avicularis relating to small birds
avium of birds
bufonius relating to toads; growing in damp places
caninus relating to dogs
caprea a nanny goat; relating to goats
caprifolius with goatlike leaves
cochleatus shaped like a snail-shell, in a spiral
colombinus like a dove
colubrinus snakelike
crista-galli a cockerel's comb
crus-galli a cockerel's spur
dens-canis a dog's tooth
dracocephalus with a dragon 's head
echiniformis shaped like a hedgehog or sea-urchin (*echinus*)
elephantipes like an elephant's foot

erinaceus like a hedgehog
formicarius attractive to ants; of ants
gruinus like a crane
lagocephalus with a head like a hare
lagopus hare-footed
leonensis like a lion
leonurus like a lion's tail
lupulus a small wolf
meleagris a guinea-fowl (spotted)
muricatus rough with sharp points (like the purple mollusc *murex*)
ophiocarpus with snakelike fruits
ornithopodus like a bird's foot
ovinus for sheep
papilio butterfly
pardalinus like a leopard (spotted)
perdicarius of partridges
psittacinus like a parrot
scolopendrius millipede
scorpioides like a scorpion
sphegodes like a wasp
tragophyllus with goatlike leaves
urophyllus with leaves like an ox's tail
ursinus bearlike
vulpinus relating to foxes

Lathyrus from Greek *lathyros,* pea or pulse

Lavandula from Latin *lavare,* to wash

Leptospermum from Greek *leptos,* slender, and *sperma,* seed

Leucanthemum from Greek *leukos,* white, and *anthemon,* flower

Lewisia named for Captain Meriwether Lewis (1774–1809), explorer in North America

Libertia named for Marie Libert (1782–1863), Belgian botanist

Ligularia from Latin *ligula,* strap

Limnanthes from Greek *limne,* marsh, and *anthos,* flower

Linnaea named for Carl Linnaeus (see page 123)

Liriope in Greek mythology, a nymph who was the mother of Narcissus

Lithospermum from Greek *lithos,* stone, and *sperma,* seed

Lobelia named for Mathias de l'Obel (1538–1616), Flemish botanist

Lonicera named for Adam Lonitzer (1528–1586), German botanist

Lunaria from Latin *luna,* the moon

leichtlinii named for the German plantsman Max Leichtlin (1831–1910)

Camassia leichtlinii

lemoinei named for the French plant breeder and nurseryman Victor Lemoine (1823–1911) and his son Emile (1862–1942)

Philadelphus × *lemoinei*

leonurus like a lion's tail

Leonotis leonurus

leonensis like a lion

Penstemon leonensis

lepido- scaly

lepto- slender (see below)

leptophyllus with slender leaves

leptorrhizus with slender rhizomes

Epimedium leptorrhizum

leucanthus white-flowered

Lilium leucanthum

leuco- white

leucocarpus with white fruit

Callicarpa japonica 'Leucocarpa'

leucodermis with white skin or bark

Acer saccharum subsp. *leucoderme*

leuconeurus white-veined

leucophaeus light grey

Digitalis leucophaea

leylandii named after C.J. Leyland, a 19th-century owner of Haggerston Hall in Northumberland, where the popular hybrid hedging conifer was bred

× *Cupressocyparis leylandii*

COMMON DESCRIPTIVE TERMS

libani, libanotis, libanoticus

Lebanese; from Mount Lebanon

Cedrus libani

Cistus libanotis

Cyclamen libanoticum

Geranium libani

Puschkinia scilloides var. *libanotica*

Labrador tea *Ledum groenlandicum*
Lad's love *Artemisia abrotanum*
Lady fern *Athyrium filix-femina*
Lady's mantle *Alchemilla mollis*
Lady's smock *Cardamine pratensis*
Ladybird poppy *Papaver commutatum*
Lamb's ears, Lamb's lugs, Lamb's tails
  Stachys byzantina
Larch *Larix*
Larkspur *Consolida ambigua*
Laurel, bay *Laurus nobilis*
Laurel, cherry *Prunus laurocerasus*
Laurel, Portugal *Prunus lusitanica*
Laurel, spotted *Aucuba japonica*
Laurustinus *Viburnum tinus*
Lavender *Lavandula*
Lavender cotton *Santolina*
Lawson cypress *Chamaecyparis
  lawsoniana*
Leadwort *Plumbago*
Lemon balm *Melissa officinalis*
Lemon verbena *Aloysia triphylla*
Lent lily *Narcissus pseudonarcissus*
Lenten rose *Helleborus orientalis;
  Helleborus × hybridus*
Leopard's bane *Doronicum orientale*
Leyland cypress   *× Cupressocyparis
  leylandii*
Lilac *Syringa vulgaris*
Lily *Lilium*
Lily-of-the-valley *Convallaria majalis*
Lilyturf *Liriope*; *Ophiopogon*
Lime, Linden *Tilia*
Ling *Calluna vulgaris*
Lobster claw *Clianthus puniceus*
Locust *Robinia*
Lombardy poplar *Populus nigra* 'Italica'
London pride *Saxifraga × urbium*

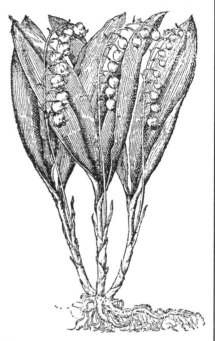

LILY-OF-THE-VALLEY

Looking-glass plant *Coprosma repens*
Loquat *Eriobotrya japonica*
Lord Anson's blue pea *Lathyrus
  nervosus*
Lords and ladies *Arum italicum*
Lotus *Nelumbo*
Love-grass *Eragrostis*
Love-in-a-mist *Nigella damascena*
Love-lies-bleeding *Amaranthus
  caudatus*
Lucerne *Medicago sativa*
Lucombe oak *Quercus × hispanica
  *'Lucombeana'
Lungwort *Pulmonaria*
Lupin *Lupinus*

WHAT'S IN A NAME?

PLANT USES

catharticus purgative, purifying
coronarius used for wreaths or garlands
domestica domesticated
dysentericus for the treatment of dysentery
edulis edible
emeticus emetic, causing vomiting
esculentus edible
febrifugus driving out fever
ferax fruit-bearing, fruitful, fertile
fructifer, frugifer fruit-bearing
frumentaceus grain-bearing
fullonum of fullers
funebris funereal; of cemeteries
medicus medicinal
officinalis sold in (apothecaries') shops; used in medicine
oleraceus relating to vegetables and herbs; of the kitchen garden (also *holeraceus*)
papyriferus paper-bearing
pulegius, pulegioides flea-repelling
ritualis ritual, of ceremonies
sativus sown, planted, cultivated
somnifer sleep-bringing
tinctorius used in dyeing
topiarius of or for topiary
torminalis curing gripe or colic
trachelius curing throat ailments
tranquillans calming
usitatissimus very useful
utilis useful
vesicarius relating to the bladder (or a remedy for bladder ailments)
vinifer wine-producing
vomitorius causing vomiting
vulnerarius of wounds; for wounds

liburnicus from the Croatian coast
Asphodeline liburnica
lichiangensis from the area around Lijiang in Yunnan Province, China
Lysimachia lichiangensis
ligustrinus like *Ligustrum* (privet)
Eupatorium ligustrinum
lilacinus lilac-coloured
Primula vulgaris 'Lilacina Plena'
liliago lily-like
Anthericum liliago
liliiflorus lily-flowered
Magnolia liliiflora
limosus of marshes
linarioides like *Linaria* (toadflax)
Penstemon linarioides
lindheimeri named for Ferdinand Lindheimer (1802–1879), a German botanist in Texas
Gaura lindheimeri
linearifolius with linear leaves
Cotoneaster linearifolius
linearis linear, narrow
Callistemon linearis
lineatus with lines, in a straight line
Convolvulus lineatus

COMMON DESCRIPTIVE TERMS

linifolius with leaves like *Linum* (flax)
Anagallis monellii subsp. *linifolia*
Erysimum linifolium
Lathyrus linifolius
Tulipa linifolia

linnaeoides named for the Swedish taxonomist Carl Linnaeus (1707–1778), the father of botanical classification and naming
   Lobelia linnaeoides

lissospermus with smooth seeds
   Hakea lissosperma

litho- stone (see below)

lithophilus stone-loving, growing in stony places
   Paeonia tenuifolia subsp. *lithophila*

lithospermus with hard seeds, like stones

littoralis of the seashore
   Griselinia littoralis

lividus leaden blue-grey
   Helleborus lividus

lizei named for the nursery of Lizé Frères in Nantes, France
   × *Fatshedera lizei*

lobatus lobed
   Ipomoea lobata

lobularis, lobulatus with small lobes
   Narcissus pseudonarcissus 'Lobularis'

locellatus divided into subsidiary compartments

locularis with cavities or chambers

lomariifolius with leaves like *Blechnum* (ferns formerly named *Lomaria*)
   Mahonia lomariifolia

longaevus long-lived
   Furcraea longaeva

CULTIVAR NAMES
Strange but True
L

IRIS ENSATA 'LAUGHING LION'

FUCHSIA 'LECHLADE CHINAMAN'

PULMONARIA SACCHARATA 'LEOPARD'

LARIX DECIDUA 'LITTLE BOGLE'

HEMEROCALLIS 'LITTLE BUGGER'

PENNISETUM ALOPECUROIDES 'LITTLE BUNNY'

HEMEROCALLIS 'LITTLE FAT DAZZLER'

PENNISETUM ALOPECUROIDES 'LITTLE HONEY'

MISCANTHUS SINENSIS 'LITTLE KITTEN'

LEWISIA COTYLEDON 'LITTLE PLUM'

OPHIOPOGON PLANISCAPUS 'LITTLE TABBY'

NEPETA RACEMOSA 'LITTLE TITCH'

HEMEROCALLIS 'LITTLE WART'

SEMPERVIVUM 'LIVELY BUG'

IRIS 'LOOP THE LOOP'

FUCHSIA 'LOTTIE HOBBY'

CROCOSMIA 'LUCIFER'

ROSA LUCKY DUCK ('DICZEST')

---

longibarbatus with a long beard

longibracteatus with long bracts
   Ornithogalum longibracteatum

longicaudatus long-tailed
   Polypodium glycyrrhiza 'Longicaudatum'

longiflorus with long flowers
   Lilium longiflorum

COMMON DESCRIPTIVE TERMS

longifolius long-leaved
Mentha longifolia
Morina longifolia
Pulmonaria longifolia

longipes with a long stalk
Nepeta longipes
longipetalus with long petals
Lewisia longipetala

longissimus very long
Hosta longissima
lophanthus with crested flowers
Paraserianthes lophantha
lucidus shining, clear, transparent
Ligustrum lucidum
ludlowii named for Frank Ludlow
(1885–1972), a British teacher and
plant collector in Tibet
Paeonia delavayi var. *ludlowii*

CULTIVAR NAMES
Natural Features

MALUS 'ADIRONDACK'

FUCHSIA 'AIREDALE'

JUNIPERUS CHINENSIS 'BLUE ALPS'

ALSTROEMERIA 'GLORY OF THE ANDES'

IRIS 'RIVER AVON'

GRISELINIA LITTORALIS 'BANTRY BAY'

HELIANTHEMUM 'BEN MORE' (ALSO
'BEN FHADA', 'BEN LEDI' ETC.)

HYACINTHUS ORIENTALIS 'BEN NEVIS'

HYDRANGEA MACROPHYLLA 'BODENSEE'

GERANIUM × *OXONIANUM*
'BRECKLAND SUNSET'

GENTIANA 'CAIRNGORM'

HELIANTHEMUM 'CHEVIOT'

IRIS 'DERWENTWATER'

ERYSIMUM 'COTSWOLD GEM'

IRIS 'CLIFFS OF DOVER'

BUDDLEJA DAVIDII 'DARTMOOR'

PULMONARIA LONGIFOLIA 'DORDOGNE'

ASTILBE 'ETNA'

HOSTA 'MOUNT EVEREST'

PHLOX PANICULATA 'MOUNT FUJI'

HEBE 'GREAT ORME'

DIANTHUS 'HAYTOR WHITE'

POLYGONATUM VERTICILLATUM
'HIMALAYAN GIANT'

IRIS 'KANGCHENJUNGA'

PENSTEMON KILIMANJARO ('YAJARO')

DELPHINIUM 'LOCH LEVEN'

PHILADELPHUS 'MONT BLANC'

HEMEROCALLIS 'NILE CRANE'

IRIS 'ORINOCO FLOW'

HOSTA 'PACIFIC BLUE EDGER'

HELLEBORUS ARGUTIFOLIUS
'PACIFIC FROST'

GERANIUM MACRORRHIZUM 'PINDUS'

MIMULUS 'POPACATAPETL'

NARCISSUS 'PURBECK'

PELARGONIUM 'RIO GRANDE'

STACHYS OFFICINALIS 'SAHARAN PINK'

GERANIUM × *OXONIANUM* 'SHERWOOD'

PELARGONIUM 'SOLENT WAVES'

GENTIANA 'STRATHMORE'

LYCHNIS × *ARKWRIGHTII* 'VESUVIUS'

ludoviciana from Louisiana, USA
  Artemisia ludoviciana
lunulatus shaped like a small crescent
  moon
  Helianthemum lunulatum
lupulus a small wolf
  Humulus lupulus
luridus dirty yellow
  Carex lurida

COMMON DESCRIPTIVE TERMS
lusitanicus from Portugal (the
  Roman province of Lusitania)
Echium lusitanicum
Fritillaria lusitanica
Lavandula stoechas subsp. *lusitanica*
Prunus lusitanica

---

luteolus yellowish
  Primula luteola
luteovenosus with yellow veins
  Hydrangea luteovenosa
lutetianus from Paris
  Circaea lutetiana (enchanter's
  nightshade)

COMMON DESCRIPTIVE TERMS
luteus yellow
Asphodeline lutea
Corydalis lutea
Digitalis lutea
Fritillaria imperialis 'Maxima Lutea'
Gentiana lutea
Mimulus luteus

---

lyallii named for David Lyall
  (1817–1895), a Scottish naval
  surgeon and naturalist who
  collected plants in New Zealand
  Hoheria lyallii
lycoctonus from *lykos* (wolf) and
  ktonos (murder)
  Aconitum lycoctonum (wolfsbane)
lydius from Lydia in Asia Minor
  Genista lydia
lyonii named for the Scottish
  gardener and botanist John Lyon
  (1765–1814), who collected plants
  in North America
  Chelone lyonii
lyratus lyre-shaped
lysolepis with loose scales

CIRCAEA LUTETIANA

M

maackii named for the Russian naturalist and traveller Richard Maack (1825–1886)
   Prunus maackii
macedonicus Macedonian
   Knautia macedonica
macranthus with large flowers
   Exochorda × *macrantha*
macro, macra large
   Hakonechloa macra
macrocalyx with a large calyx
   Cyananthus macrocalyx

WHAT'S IN A NAME?

COLOUR: GREY AND SILVER

argentatus silvered
argenteus silvery, lustrous
argyrotrichon silver-haired
canescens rather hoary, whitish grey
cinereus, cineraceus ash grey
ferreus iron grey
fulmineus lightning-coloured
fumeus, fumidus, fumosus smoky grey
glaucescens rather glaucous, becoming glaucous
glaucus glaucous
griseus pure pearly grey
incanus pale, whitish grey; hoary
leucophaeus light grey
murinus mousy grey
plumbeus leaden
schistaceus slate grey

COMMON DESCRIPTIVE TERMS
macrocarpus with large fruit
   Cupressus macrocarpa
   Hebe macrocarpa
   Muscari macrocarpum
   Oenothera macrocarpa
   Quercus macrocarpa

---

macrocephalus with a large head
   Carex macrocephala
macrodontus with large teeth
   Olearia macrodonta
macropetalus with large petals
   Clematis macropetala

COMMON DESCRIPTIVE TERMS
macrophyllus with large leaves
   Brunnera macrophylla
   Cornus macrophylla
   Hydrangea macrophylla
   Magnolia macrophylla
   Pachyphragma macrophyllum
   Persicaria macrophylla
   Primula macrophylla

---

macropodus with a large stalk
   Daphniphyllum himalaense subsp. *macropodum*
macrorrhizus with large roots
   Geranium macrorrhizum
macrostachyus with a large spike
   Pennisetum macrostachyum
macrostylus with a large style
   Geranium macrostylum

CULTIVAR NAMES
Personal Names
M

SEMPERVIVUM 'MADELEINE'
CANNA 'MAGGIE'
FUCHSIA 'AMAZING MAISIE'
FUCHSIA 'MARCIA'
FUCHSIA 'MARGARET'
HELENIUM 'MARGOT'
CORREA 'MARIAN'S MARVEL'
HEBE 'BABY MARIE'
ACHILLEA 'MARIE ANN'
TULIPA 'MARIETTE'
TULIPA 'MARILYN'
ARBUTUS 'MARINA'
ABUTILON 'MARION'
PRUNUS DOMESTICA (PLUM) 'MARJORIE'S SEEDLING'
PRIMULA AURICULA 'MARK'
OSTEOSPERMUM 'SUNNY MARTHA'
VIOLA 'MARTIN'
ACHILLEA 'MARTINA'
FUCHSIA 'MARY'
HEDERA HELIX 'MATHILDE'

CLEMATIS 'MAUREEN'
NARCISSUS 'MAX'
PELARGONIUM 'MAXINE'
SISYRINCHIUM STRIATUM 'AUNT MAY'
CLEMATIS 'MAYLEEN'
ROSA 'MEG'
RHODODENDRON 'MEGAN'
HEDERA HELIX 'MELANIE'
PARAHEBE 'MERVYN'
OLEARIA × *SCILLONIENSIS* 'MASTER MICHAEL'
PRIMULA AURICULA 'MILLICENT'
RHODODENDRON 'MIMI'
HYDRANGEA SERRATA 'MIRANDA'
PELARGONIUM 'SWEET MIRIAM'
ALSTROEMERIA PRINCESS MONICA 'STAPRIMON'
PELARGONIUM 'MORWENNA'
FUCHSIA 'MOYRA'
FUCHSIA 'MURIEL'

COMMON DESCRIPTIVE TERMS
maculatus, maculifer spotted, blotched
Arum maculatum
Dactylorhiza maculata
Elaeagnus pungens 'Maculata'
Geranium maculatum
Lamium maculatum
Phlox maculata
Rhododendron maculiferum

madagascariensis from the island of Madagascar
Buddleja madagascariensis

COMMON DESCRIPTIVE TERMS
maderensis from Madeira
Argyranthemum maderense
Geranium maderense
Hedera maderensis
Ilex × *altaclerensis* 'Maderensis'

CULTIVAR NAMES
Foreign Expressions
M

MAIGRÜN (*German*) MAY GREEN
(*LONICERA NITIDA* 'MAIGRÜN')

MAINACHT (*German*) MAY NIGHT
(*SALVIA* × *SYLVESTRIS* 'MAINACHT')

MAISCHNEE (*German*) MAY SNOW
(*PHLOX SUBULATA* 'MAISCHNEE')

MANTEAU D'HERMINE (*French*)
ERMINE CLOAK
(*PHILADELPHUS* 'MANTEAU
D'HERMINE')

MARÉE D'OR (*French*) GOLD TIDE
(*FORSYTHIA* MARÉE D'OR
('COURTASOL')

MATKA TERESA (*Polish*)
MOTHER TERESA
(*CLEMATIS* 'MATKA TERESA')

MILCHSTRASSE (*German*)
THE MILKY WAY
(*PULMONARIA* 'MILCHSTRASSE')

MIYAMA-YAE-MURASAKI (*Japanese*)
MOUNTAIN DOUBLE PURPLE
(*HYDRANGEA SERRATA*
'MIYAMA-YAE-MURASAKI')

MOORHEXE (*German*)
MOORLAND WITCH
(*MOLINIA CAERULEA* SUBSP. *CAERULEA*
'MOORHEXE')

MORGENRÖTE (*German*)
MORNING RED
(*BERGENIA* 'MORGENRÖTE')

MÖWE (*German*) SEAGULL
(*HYDRANGEA MACROPHYLLA* 'MÖWE')

COMMON DESCRIPTIVE TERMS

magellanicus from the southern
tip of South America, around the
Straits of Magellan
Elymus magellanicus
Fuchsia magellanica
Gunnera magellanica
Philesia magellanica

---

magnificus magnificent, splendid
Geranium × *magnificum*
magniflorus large-flowered
Geranium magniflorum
magnus large
Ornithogalum magnum
majalis of May, flowering in May
Convallaria majalis

COMMON DESCRIPTIVE TERMS

major, majus greater
Ammi majus
Antirrhinum majus
Astrantia major
Cerinthe major
Fothergilla major
Lagarosiphon major
Melianthus major
Tropaeolum majus
Vinca major

---

makinoi named for the Japanese
botanist Tomitaro Makino
(1863–1957)
Rhododendron makinoi

malacoides soft, supple
    Primula malacoides
malviflorus with flowers like *Malva*
    (mallow)
    Sidalcea malviflora

COMMON DESCRIPTIVE TERMS
mandschuricus, mandshuricus,
  manshuriensis from Manchuria
Acer mandschuricum
Betula mandshurica
Clematis mandschurica
Malus baccata var. *mandshurica*
Prunus mandshurica

---

manicatus long-sleeved
    Gunnera manicata
marantinus like *Maranta* (arrowroot)
    Globba marantina
margaritaceus pearly
    Anaphalis margaritacea (pearl
    everlasting)

COMMON DESCRIPTIVE TERMS
marginalis, marginatus with a
  distinct margin, edge or border
Buxus sempervirens 'Marginata'
Dryopteris marginalis
Ilex aquifolium 'Argentea Marginata'
Luzula sylvatica 'Marginata'
Primula marginata
Sambucus nigra 'Marginata'
Teucrium scorodonia 'Crispum
  Marginatum'

WHAT'S IN A NAME?

SIZE

altissimus very tall
altus tall
decumanus very large
diminutus small, diminished
elatus tall; *elatior, elatius* taller
excelsus lofty, high
exiguus small, weak
giganteus giant, gigantic, very large
gigas giant
grandis big
magnus large
major, majus greater
maximus largest, very large
mega-, megalo- large
minimus smallest, very small
minor smaller
minutissimus most minute, very
  minute
nanus dwarf
parvi- small
procerus high, tall
pumilus dwarf
pygmaeus pygmy, dwarf, very small
vescus small, feeble

---

marianus named for the Virgin Mary
  Silybum marianum

COMMON DESCRIPTIVE TERMS
mariesii named for the British plant
collector Charles Maries (1851–
1902)
Ilex crenata 'Mariesii'
Platycodon grandiflorus 'Mariesii'
Viburnum plicatum 'Mariesii

CULTIVAR NAMES
Strange but True
M

IRIS 'MAKING EYES'

IRIS 'MARMALADE SKIES'

KNAUTIA MACEDONICA 'MARS MIDGET'

POTENTILLA FRUTICOSA 'MEDICINE WHEEL MOUNTAIN'

IRIS 'MESMERIZER'

FUCHSIA 'MICROCHIP'

VERONICA CHAMAEDRYS 'MIFFY BRUTE'

GALANTHUS 'MIGHTY ATOM'

HEMEROCALLIS 'MILANESE MANGO'

PINUS MUGO 'MINIKIN'

ROSA 'MINNEHAHA'

JUNIPERUS × *PFITZERIANA* 'MINT JULEP'

PAEONIA LACTIFLORA 'MISCHIEF'

ROSA MISS FLIPPINS ('TUCKFLIP')

LAVANDULA ANGUSTIFOLIA MISS MUFFET ('SCHOLMIS')

PRIMULA AURICULA 'MR "A"'

NARCISSUS 'MITE'

PENSTEMON 'MODESTY'

PELARGONIUM 'THE MOLE'

PINUS MUGO 'MOPS' (AND 'MINI MOPS')

CALLUNA VULGARIS 'MOUSEHOLE'

HOSTA 'MR BIG'

GERANIUM × *MONACENSE* VAR. *MONACENSE* 'MULDOON'

HEMEROCALLIS 'MUMBO JUMBO'

BEGONIA 'MUNCHKIN'

LUPINUS 'MY CASTLE'

marilandicus from Maryland, USA
Spigelia marilandica 'Wisley Jester'

COMMON DESCRIPTIVE TERMS

maritimus coastal, growing by the sea
Armeria maritima (thrift)
Crambe maritima (seakale)
Eryngium maritimum (sea holly)
Lavatera maritima
Lobularia maritima (sweet alyssum)

CRAMBE MARITIMA

marmorarius of marble
marmoratus marbled
 Arum italicum subsp. *italicum* 'Marmoratum'
mas, masculus male, masculine
 Cornus mas
 Paeonia mascula

PLANT HABIT

apodus without a foot or stalk; sessile

arborescens growing to be a tree; woody

arboreus treelike, woody

arbusculus like a small tree

caespitosus, cespitosus clump-forming

cernuus drooping, downturned

columnaris columnar

contortus twisted, contorted

decumbens decumbent; prostrate but with upright tips

dependens hanging down

dilatatus spread out, expanded

divaricatus spreading, straggling

dumosus bushy

effusus spreading, straggly

erectus erect, upright

fastigiatus fastigiate; with an upright habit

fruticosus, frutescens, fruticans shrubby

gracilis slender

ortho- straight, upright (see below)

patens spreading

patulus spreading

pendulus pendulous, hanging

procumbens prostrate

procurrens spreading, running

prostratus prostrate

ramosus branched

ramulosus twiggy

rectus upright

repens, reptans creeping

scandens climbing

sessilis sessile, stalkless

sobolifer with creeping stems that form roots

socialis growing in colonies

spiralis spiral

stans upright

sten-, steno- narrow

stolonifer with stolons (rooting runners)

stragulus, stragulatus mat-forming

strictus upright, erect, tight

suffruticosus somewhat shrubby

suspensus hanging

tenuis thin, fine, slender

tenuissimus very thin

tortuosus, tortus, tortilis tortuous, winding

verticillatus verticillate; arranged in whorls

viminalis, vimineus with long, thin shoots; like osiers

virgatus twiggy

volubilis winding, revolving

---

matronalis named for *Matronalia*, a Roman festival on 1 March

Hesperis matronalis

maxillaris relating to jaws

maximowiczianus, maximowiczii named after a Russian botanist, Carl Maximowicz (1827–1891)

Geranium maximowiczii

COMMON DESCRIPTIVE TERMS

maximus largest, very large,

Astrantia maxima

Briza maxima

Corylus maxima

Glyceria maxima

Hepatica maxima

Rhododendron maximum

Macleaya named for Alexander Macleay (1767–1848), one-time secretary of the Linnaean Society

Magnolia named for Pierre Magnol (1638–1715), French botanist

Mahonia named for Bernard M'Mahon (1775–1816), American horticulturist

Malus from Greek *melon*, apple

Matteucia named for Carlo Matteuci (1811–1868), Italian scientist

Matthiola named for Pietro Andrea Mattioli (1500–1577), Italian botanist and doctor

Melianthus from Greek *meli*, honey, and *anthos*, flower

Melissa in Greek, a bee

Miscanthus from Greek *miskos*, stem, and *anthos*, flower

Monarda named for Nicolas Monardes (1493–1588), Spanish botanist

Myosotis from Greek *mus*, mouse, and *otos*, ear

---

medicus medicinal

COMMON DESCRIPTIVE TERMS

medius mid, in the middle, intermediate

Briza media

Campanula medium

Dierama media

Mahonia × *media*

Plantago media

---

mega-, megalo- large (see below)

megacalyx with a large calyx

   Rhododendron megacalyx

megalanthus with large flowers

   Potentilla megalantha

megalocarpus with large fruits

   Sorbus megalocarpa

megalophyllus with large leaves

   Ampelopsis megalophylla

megapotamicus (literally) from the big river; from the Rio Grande

   Abutilon megapotamicum

mela-, melano- black

melanocarpus with black fruit

   Aronia melanocarpa

melanostictus spotted with black

meleagris a guinea-fowl; hence, spotted

   Fritillaria meleagris

meliodorus honey-scented

melissifolius, melissophyllus with leaves like *Melissa* (lemon balm)

   Melittis melissophyllum

melleus, mellitus sweet like honey

mellifer honey-bearing

   Euphorbia mellifera

membranaceus like skin or membrane

   Epimedium membranaceum

meniscifolius with concave leaves

menthifolius with leaves like *Mentha* (mint)

   Calamintha menthifolia

Madonna lily *Lilium candidum*
Maidenhair fern *Adiantum*
Maidenhair spleenwort *Asplenium trichomanes*
Maidenhair tree *Ginkgo biloba*
Maize *Zea mays*
Male fern *Dryopteris filix-mas*
Mallow *Lavatera; Malope; Malva*
Maltese cross *Lychnis chalcedonica*
Manchurian cherry *Prunus maackii*
Mandrake *Mandragora*
Manzanita *Arbutus; Arctostaphylos*
Maple *Acer*
Marguerite *Argyranthemum frutescens; Leucanthemum vulgare*
Marigold *Calendula; Tagetes*
Mariposa tulip *Calochortus*
Marjoram *Origanum*
Marsh marigold *Caltha palustris*
Marvel of Peru *Mirabilis jalapa*
Masterwort *Astrantia*
Mastic tree *Pistacia lentiscus*
May *Crataegus monogyna; Crataegus laevigata*
Meadow cranesbill *Geranium pratense*
Meadow foxtail *Alopecurus pratensis*
Meadow rue *Thalictrum*
Meadow saffron *Colchicum*
Meadowsweet *Filipendula ulmaria*
Medick *Medicago*
Medlar *Mespilus germanica*
Melick grass *Melica*
Mexican breadfruit *Monstera deliciosa*
Mexican hat plant *Kalanchoe daigremontiana*
Mexican orange blossom *Choisya ternata*
Mezereon *Daphne mezereum*

Michaelmas daisy *Aster*
Midland hawthorn *Crataegus laevigata*
Mignonette *Reseda odorata*
Mile-a-minute vine *Fallopia baldschuanica*
Milk thistle *Silybum marianum*
Milkweed *Asclepias*
Milkwort *Polygala*
Mimosa *Acacia dealbata*
Mind-your-own-business *Soleirolia soleirolii*
Mint *Mentha*
Mint bush *Prostanthera*
Mintleaf *Plectranthus madagascariensis*
Miss Willmott's ghost *Eryngium giganteum*
Missey-moosey *Sorbus americana*
Mistletoe *Viscum album*
Mock orange *Philadelphus*
Mole plant *Euphorbia lathyris*
Monkey flower *Mimulus*
Monkey-puzzle tree *Araucaria araucana*
Monkshood *Aconitum*
Montbretia *Crocosmia*
Moon daisy *Leucanthemum vulgare*
Morning glory *Ipomoea tricolor*
Moss rose *Rosa* × *centifolia* 'Muscosa'
Mount Etna broom *Genista aetnensis*
Mountain ash *Sorbus aucuparia*
Mourning widow *Geranium phaeum*
Moutan *Paeonia suffruticosa*
Mulberry *Morus*
Mullein *Verbascum*
Musk mallow *Malva moschata*
Musk rose *Rosa moschata*
Myrobalan *Prunus cerasifera*
Myrtle *Myrtus communis*
Monterey cypress *Cupressus macrocarpa*

CULTIVAR NAMES
Plants for all Seasons

SPRING

CALLUNA VULGARIS 'SPRING CREAM'

CHAMAECYPARIS LAWSONIANA 'SPRINGTIME'

CLEMATIS 'FRAGRANT SPRING'

CORNUS FLORIDA 'SPRING SONG'

DICENTRA 'SPRING MORNING'

LATHYRUS VERNUS 'SPRING MELODY'

PAPAVER ORIENTALE 'SPRINGTIME'

PELARGONIUM 'FLOWER OF SPRING'

PRIMULA AURICULA 'SPRING MEADOWS'

PRUNUS CERASIFERA 'SPRING GLOW'

TIARELLA 'SPRING SYMPHONY'

SUMMER

CANNA 'SUMMER GOLD'

CHAMAECYPARIS LAWSONIANA 'SUMMER SNOW'

DIGITALIS 'SALTWOOD SUMMER'

GERANIUM PYRENAICUM 'SUMMER SNOW'

GERANIUM SUMMER SKIES ('GERNIC')

GERANIUM × *OXONIANUM* 'SUMMER SURPRISE'

HEBE 'MIDSUMMER BEAUTY'

HEBE ODORA 'SUMMER FROST'

HEMEROCALLIS 'SUMMER WINE'

LAVATERA OLBIA 'SUMMER KISSES'

LEUCANTHEMUM × *SUPERBUM* 'SUMMER SNOWBALL'

OENOTHERA 'SUMMER SUN'

PHYSOSTEGIA VIRGINIANA 'SUMMER SNOW'

RHODODENDRON 'HIGH SUMMER'

ROSA SUMMER WINE ('KORIZONT')

AUTUMN

ASTER NOVI-BELGII 'AUTUMN GLORY'

COLCHICUM 'AUTUMN QUEEN'

CAMPSIS RADICANS 'INDIAN SUMMER'

DAHLIA 'AUTUMN LUSTRE'

GINKGO BILOBA 'AUTUMN GOLD'

HEBE 'AUTUMN GLORY'

HELENIUM 'AUTUMN LOLLIPOP'

HEMEROCALLIS 'AUTUMN RED'

LEPTOSPERMUM SCOPARIUM 'AUTUMN GLORY'

MISCANTHUS SINENSIS 'AUTUMN LIGHT'

ROSA 'AUTUMN DELIGHT'

RUBUS IDAEUS (RASPBERRY) 'AUTUMN BLISS'

SORBUS AUTUMN SPIRES ('FLANROCK')

WINTER

BUXUS SINICA VAR. *INSULARIS* 'WINTER GEM'

CORNUS SANGUINEA 'MIDWINTER FIRE'

ERICA CARNEA 'WINTER BEAUTY'

EUONYMUS HAMILTONIANUS 'WINTER GLORY'

GAULTHERIA MUCRONATA 'WINTERTIME'

HEBE 'WINTER GLOW'

ILEX VERTICILLATA 'WINTER RED'

LEPTOSPERMUM SCOPARIUM 'WINTER CHEER'

LONICERA × *PURPUSII* 'WINTER BEAUTY'

MAHONIA × *MEDIA* 'WINTER SUN'

MALUS DOMESTICA (APPLE) 'WINTER BANANA'

menziesii named for the Scottish plant collector, surgeon and naturalist Archibald Menzies (1754–1842)
Arbutus menziesii
Nothofagus menziesii
Pseudotsuga menziesii (Douglas fir)
Sanguisorba menziesii
Tolmiea menziesii

meridionalis (flowering at) midday, noon
Globularia meridionalis
meserveae named for Kathleen Meserve, who in the 1950s bred hollies on Long Island, New York
Ilex × *meserveae*
messanensis from Messina, Sicily
Fritillaria messanensis
metallicus metallic
Begonia metallica

mexicanus Mexican
Agastache mexicana
Penstemon mexicanus
Philadelphus mexicanus
Quercus mexicana

michiganensis from Michigan, USA
Lilium michiganense
micr-, micro- small

micranthus with small flowers
Acer micranthum
Colchicum micranthum
Dichondra micrantha
Heuchera micrantha
Rhododendron micranthum

microcarpus with small fruits
Macleaya microcarpa
microcephalus with a small head
Persicaria microcephala
microlepis with small scales
Dianthus microlepis

microphyllus with small leaves
Azara microphylla
Buxus microphylla
Cotoneaster microphyllus
Euonymus japonicus 'Microphyllus'
Fuchsia microphylla
Helichrysum microphyllum
Myrtus communis 'Microphylla Variegata'
Origanum microphyllum
Philadelphus microphyllus
Sophora microphylla

middendorffianus, middendorfii named for Alexander von Middendorff (1815–1894), a plant collector and zoologist in Siberia
Sedum middendorffianum

WHAT'S IN A NAME?

PLANTS OF DISTINCTION

admirabilis admirable, wonderful
amabilis lovely
amoenus lovely, pleasant
augustus venerable, majestic
basilicus princely, royal
benedictus well spoken of; blessed
concinnus neat
decoratus, decorus decorative
elegans elegant
elegantissimus very elegant
erromenus vigorous, healthy
eucharis pleasing, agreeable
eudoxus of good repute
eximius distinguished, extraordinary
facetus choice, fine
formosissimus very beautiful
formosus finely formed, handsome, beautiful
fortis strong, vigorous
generosus noble, eminent
gloriosus glorious
imperialis imperial, showy
insignis distinguished
jucundus pleasant, delightful
magnificus magnificent, splendid
mirabilis marvellous
nobilis noble, renowned
ornatus showy
praestans distinguished, excellent
princeps princely, most distinguished
pulcher, pulchellus beautiful, pretty
regalis royal; fit for a king
robustus robust, strong
speciosus showy, splendid
spectabilis spectacular
superbus superb
validus strong, robust, powerful
venustus lovely, charming, pleasing

miliaceus like millet
millefolius, millefoliatus (literally) with a thousand leaves
  Achillea millefolium (yarrow)
miniatus the colour of red-lead or cinnabar
  Clivia miniata
minimus smallest, very small
  Ocimum minimum (bush basil)

COMMON DESCRIPTIVE TERMS

minor smaller
Hemerocallis minor
Hosta minor
Rhinanthus minor (yellow rattle)
Vinca minor

---

minutiflorus with very small flowers
  Origanum minutiflorum
minutissimus most minute, very minute
  Athyrium filix-femina 'Minutissimum'
minutus minute, very small
  Fritillaria minuta
mirabilis marvellous
  Cyclamen mirabile
mischtschenkoanus named for P.I. Misczenko (1869–1938), a Russian botanist
  Scilla mischtschenkoana
missouriensis from Missouri or the Missouri River, USA
  Iris missouriensis

mitriformis shaped like a cap or hat
   Aloe mitriformis
mlokosewitschii named for Ludwik
   Mlokosiewicz (1831–1909),
   a Polish naturalist
   Paeonia mlokosewitschii
modestus modest

COMMON DESCRIPTIVE TERMS

mollis soft
Acanthus mollis
Alchemilla mollis
Hamamelis mollis
Helianthus mollis
Holcus mollis
Olearia × *mollis*
Pulmonaria mollis

mollissimus very soft
   Passiflora mollissima
monacensis of or from Munich,
   Germany
   Geranium × *monacense*
mongolicus Mongolian
   Tilia mongolica
mono- single
monocephalus with one head
   Anaphalis nepalensis var.
   monocephala
monogynus with one pistil
   Crataegus monogyna (hawthorn)
monopetalus with one petal
monostictus with one spot
   Galanthus elwesii var. *monostictus*

COMMON DESCRIPTIVE TERMS

monspeliensis, monspessulanus
   from Montpellier, France
Acer monspessulanum
Aphyllanthes monspeliensis
Cistus monspeliensis
Dianthus monspessulanus

CULTIVAR NAMES
Colours: Orange

ALSTROEMERIA 'ORANGE GLORY'
BERBERIS LINEARIFOLIA
'ORANGE KING'
CANNA 'ORANGE PERFECTION'
ERICA × *STUARTII* 'IRISH ORANGE'
ERYSIMUM 'ORANGE FLAME'
GEUM 'PRINCE OF ORANGE'
HAMAMELIS × *INTERMEDIA*
'ORANGE PEEL'
HEDYCHIUM DENSIFLORUM
'ASSAM ORANGE'
HELENIUM 'CHIPPERFIELD ORANGE'
IRIS 'ORANGE CAPER'
LILIUM 'ORANGE PIXIE'
MALUS DOMESTICA (APPLE)
'ELLISON'S ORANGE'
MIMULUS 'HIGHLAND ORANGE'
OSTEOSPERMUM ORANGE SYMPHONY
('SEIMORA')
PHLOX PANICULATA
'PRINCE OF ORANGE'
POTENTILLA FRUTICOSA
'HOPLEYS ORANGE'
PYRACANTHA 'ORANGE GLOW'
RHODODENDRON 'ORANGE BEAUTY'
TULIPA 'ORANGE EMPEROR'

WHAT'S IN A NAME?

SCENT AND TASTE

agathosmus smelling good, fragrant
aromaticus fragrant, aromatic, spicy
blandus mild, not bitter; pleasing
citriodorus lemon-scented
deliciosus delicious
dulcis sweet, pleasant
foetidus evil-smelling
fragrans fragrant
glycyrrhizus like liquorice
graveolens strongly scented
inodorus scentless
meliodorus honey-scented
moschatus musk-scented
odoratus, odorus scented, fragrant
oxycarpus with sour fruits
phu rotten-smelling; acrid
saccharatus, saccharinus sweet, sugary
salsillus salted
suaveolens fragrant, smelling sweetly

COMMON DESCRIPTIVE TERMS

montanus of mountains
Arenaria montana
Centaurea montana
Clematis montana
Fritillaria montana
Geum montanum
Lathyrus montanus

---

montevidensis from Montevideo
  Lantana montevidensis
monticola a mountain-dweller
  Pinus monticola
morifolius mulberry-leaved

COMMON DESCRIPTIVE TERMS

moschatus musk-like, musk-scented
Chelonopsis moschata
Malva moschata
Narcissus moschatus
Olearia moschata
Rosa moschata

---

moupinensis from Baoxing (Mupin)
  in Sichuan, China
  Primula moupinensis
moyesii named for an early 20th-
  century missionary in China,
  the Rev. J. Moyes
  Rosa moyesii
mucosus slimy
mucronatus pointed
  Gaultheria mucronata
mucronifolius with pointed leaves
  Androsace mucronifolia
multi- many (see below)
multiceps many-headed

COMMON DESCRIPTIVE TERMS

multifidus with many divisions,
  much-divided
Anemone multifida
Brachyscome multifida
Helleborus multifidus
Lavandula multifida

---

multiflorus with many flowers
  Polygonatum multiflorum
  (Solomon's seal)

multijugus 'many yoked together' (e.g. a leaf with many leaflets)

multiscapoideus with many scapes (bare flower stems)

    Erythronium multiscapoideum

multisectus with many cuts

munitus fortified; protected

    Polystichum munitum

muralis of walls, growing on walls

    Cymbalaria muralis

muricatus rough with sharp points: a term derived from the purple mollusc *Murex*

    Pinus muricata

murielae named by the plant hunter E.H. Wilson (see page 212) after his daughter Muriel

    Fargesia murielae

musaceus like Musa (banana)

musaicus like a mosaic

muscosus mossy, like moss

    Rosa × centifolia 'Muscosa'

musifolius with leaves like *Musa* (banana)

    Canna 'Musifolia'

mutabilis changing, changeable (e.g. in colour)

    Rosa × odorata 'Mutabilis'

myiagrus sticky (named for Myiagros, a fly-catching deity of the ancient Greeks)

myosotidifolius with leaves like *Myosotis* (forget-me-not)

    Buxus sempervirens 'Myosotidifolia'

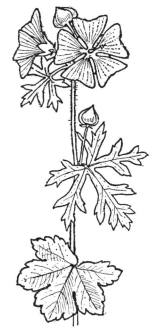

MALVA MOSCHATA

myri- many

myrsinites like myrtle

    Euphorbia myrsinites

COMMON DESCRIPTIVE TERMS

myrtifolius with leaves like *Myrtus* (myrtle)

Buxus sempervirens 'Myrtifolia'

Ilex aquifolium 'Myrtifolia'

Leptospermum myrtifolium

Polygala myrtifolia

Quercus myrtifolia

---

myrtinervius with veins like *Myrtus* (myrtle)

    Dianthus myrtinervius

N

nankingensis from Nanjing, China
 Chrysanthemum nankingense

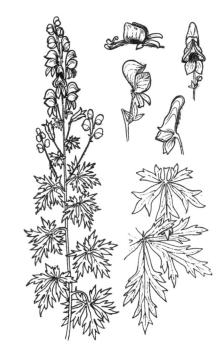

ACONITUM NAPELLUS

COMMON DESCRIPTIVE TERMS

nanus dwarf
 Betula nana
 Cryptomeria japonica 'Elegans nana'
 Larix kaempferi 'Nana'
 Lilium nanum

---

napaulensis from Nepal (see also
 nepalensis)
 Meconopsis napaulensis
napellus little turnip
 Aconitum napellus
napolitanus from Naples, Italy
 (see also *neapolitanus*)
 Daphne napolitana
narbonensis from Narbonne, France
 Linum narbonense
narcissiflorus with flowers like
 Narcissus
 Anemone narcissiflorus
neapolitanus (see also *napolitanus*)
 from Naples, Italy
 Allium neapolitanum
nebulosus like a cloud
nectarifer nectar-bearing
neglectus neglected, overlooked
 Muscari neglectum

nemorosus growing in woods
 Anemone nemorosa

COMMON DESCRIPTIVE TERMS

nepalensis from Nepal (see also
 napaulensis)
 Anaphalis nepalensis var. *monocephala*
 Lilium nepalense
 Miscanthus nepalensis
 Potentilla nepalensis
 Piptanthus nepalensis

---

nepenthoides like *Nepenthes* (pitcher
 plant)
 Arisaema nepenthoides

nephro- kidney-
nephrolepis with kidney-shaped scales

nervosus fibrous, sinewy; with
   conspicuous veins
Astelia nervosa
Cynoglossum nervosus
Lathyrus nervosus
Mahonia nervosa
Nepeta nervosa
Onopordum nervosum

CULTIVAR NAMES
Strange but True
N, O
PRIMULA AURICULA 'NEAT AND TIDY'
PELARGONIUM 'NERVOUS MABEL'
ARMERIA 'NIFTY THRIFTY'
PITTOSPORUM TENUIFOLIUM
'NUTTY'S LEPRECHAUN'
SEMPERVIVUM 'ODDITY'
DIANTHUS
'OLD MOTHER HUBBARD'
PRIMULA AURICULA
'OLD RED DUSTY MILLER'
PRIMULA AURICULA 'OLD SMOKEY'
PELARGONIUM 'OLD SPICE'
DIANTHUS 'OLD SQUARE EYES'
CAREX NIGRA 'ON-LINE'
HEBE 'ORPHAN ANNIE'
PENSTEMON 'OSPREY'
FUCHSIA 'OTHER FELLOW'
NARCISSUS 'OZ'

---

neso- of islands
nesophilus island-loving, growing
   on islands
Melaleuca nesophila
neurolobus with veined lobes
Lathyrus neurolobus
nevadensis from Nevada, USA, or
   from the Sierra Nevada (Spain
   or California)
Lewisia nevadensis
nicaeensis from Nice, France; or
   Nicaea (now Iznik), Turkey
Euphorbia nicaeensis
nidiformis nest-shaped
Picea abies 'Nidiformis'
nidulans (literally) nesting; partially
   enclosed
nidularius like a little nest
Phyllostachys nidularia

---

niger black
Allium nigrum
Carex niger
Helleborus niger (Christmas rose)
Hydrangea macrophylla 'Nigra'
Juglans nigra (walnut)
Morus nigra (black mulberry)
Lathyrus nigra
Phyllostachys nigra (black bamboo)
Prunus cerasifera 'Nigra'
Sambucus nigra (elder)
Veratrum nigrum
Verbascum nigrum

WHAT'S IN A NAME?

HOW MANY?

aequitrilobus with three equal lobes
bicornis with two horns
bicuspidatus with two points
bidentatus with two teeth
biflorus with two flowers
bifolius with two leaves
bilobus, bilobatus with two lobes
bisectus divided into two equal parts
biternatus twice ternate
decapetalus with ten petals
decaphyllus with ten leaves or leaflets
dichotomus dividing repeatedly in two
didymus paired, two-lobed
digynus with two styles or carpels
diphyllus two-leaved
dipterus two-winged
distachyus with two spikes
distichus in two rows
distylus with two styles
dodecandrus with 12 stamens
enneaphyllus with nine leaflets or leaves
mono- single
monocephalus with one head
monogynus with one style or carpel
monopetalus with one petal
oct-, octo- eight
octopetalus with eight petals
olig-, oligo- few
oligophyllus with few leaves
pauci- few
pauciflorus with few flowers
penta- five
pentapetalus with five petals
pentaphyllus with five leaves or leaflets
poly- many
quadrangularis, quadrangulatus with four angles
quadratus square; in fours; in four
quadridentatus with four teeth

quadrifolius with four leaves or leaflets
quinatus in fives
quinquefolius with five leaves
quinquelobatus with five lobes
quintuplinervius with five veins
septem- seven
septemfidus divided into seven
septemlobus with seven lobes
sexstylosus with six styles
ternatus in threes
terniflorus with flowers in threes
ternifolius with three leaves or leaflets
tetra- four
tetragonus with four angles
tetrandrus with four stamens
tetraphyllus with four leaves or leaflets
tetrapterus four-winged
triacanthos three-thorned
triandrus with three stamens
triangularis three-angled
trichotomum with divisions in threes
tricornis with three horns
tricuspidatus three-pointed
trifidus divided or cleft into three
triflorus three-flowered
trifolius; trifoliatus three-leaved
trifurcatus three-pronged
trilobus, trilobatus three-lobed
trinervis, triplinervis three-veined
tripartitus divided into three parts
tripetalus with three petals
triphyllus three-leaved
tripinnatus thrice pinnate
tripteris three-winged
triquetrus three-cornered
triternatus thrice ternate
uni- one
uniflorus with one flower
unifolius with one leaf

nigercors a hybrid between
Helleborus niger and *Helleborus argutifolius* (*Helleborus corsicus*)
Helleborus × *nigercors*

COMMON DESCRIPTIVE TERMS
nigrescens, nigricans blackish
Cistus × *nigricans*
Cytisus nigricans
Fuchsia nigricans
Hosta nigrescens
Ophiopogon planiscapus
'Nigrescens'

nigropunctatus black-spotted
niloticus from the River Nile
niphophilus snow-loving
Eucalyptus pauciflora subsp.
niphophila

COMMON DESCRIPTIVE TERMS
niponicus, nipponicus Japanese
Athyrium niponicum
Nipponanthemum nipponicum
Phyllodoce nipponica
Prunus nipponica
Spiraea nipponica

nitens, nitidus shining
Dianthus nitidus
Eucalyptus nitens
Fargesia nitida
Lonicera nitida
Potentilla nitida

CULTIVAR NAMES
Foreign Expressions
N, O

NACHTHIMMEL (*German*)
NIGHT SKY
(*ERIGERON* 'NACHTHIMMEL')

NACHTIGALL (*German*)
NIGHTINGALE
(*HYDRANGEA MACROPHYLLA* 'NACHTIGALL')

OCHSENBLUT (*German*)
OX BLOOD
(*PHLOX DOUGLASII* 'OCHSENBLUT')

OEIL DE POURPRE (*French*)
PURPLE EYE
(*PHILADELPHUS* 'OEIL DE POURPRE')

OISEAU BLEU (*French*)
BLUE BIRD
(*HIBISCUS SYRIACUS* 'OISEAU BLEU')

OKAME (*Japanese*)
A MASK USED IN TRADITIONAL JAPANESE COMEDY
(*PRUNUS* 'OKAME')

ORANJE BOVEN (*Dutch*)
ORANGE TOP
(*HYACINTHUS ORIENTALIS* 'ORANJE BOVEN')

ORO DI BOGLIASCO (*Italian*)
BOGLIASCO GOLD
(*HEDERA HELIX* 'ORO DI BOGLIASCO')

OSAKAZUKI (*Japanese*)
SAKE (RICE WINE) CUP
(*ACER PALMATUM* 'OSAKAZUKI')

OSHIDORI (*Japanese*)
MANDARIN DUCK
(*PRUNUS INCISA* 'OSHIDORI')

CULTIVAR NAMES
Colours: Silver

ARTEMISIA ABSINTHIUM 'LAMBROOK SILVER'

CLEMATIS 'SILVER MOON'

CORTADERIA SELLOANA 'SUNNINGDALE SILVER'

DICHONDRA MICRANTHA 'SILVER FALLS'

ELAEAGNUS 'QUICKSILVER'

EUONYMUS FORTUNEI 'SILVER QUEEN'

HEBE PIMELEOIDES 'QUICKSILVER'

HEBE 'SILVER DOLLAR'

HELLEBORUS ARGUTIFOLIUS 'SILVER LACE'

HOSTA 'SILVERY SLUGPROOF'

ILEX AQUIFOLIUM 'SILVER MILKMAID'

IRIS SIBIRICA 'SILVER EDGE'

JUNIPERUS VIRGINIANA SILVER SPREADER ('MONA')

LAMIUM MACULATUM 'BEACON SILVER'

LAVATERA TRIMESTRIS 'SILVER CUP'

LONICERA NITIDA 'SILVER BEAUTY'

NARCISSUS 'SILVER CHIMES'

OZOTHAMNUS ROSMARINIFOLIUS 'SILVER JUBILEE'

PHLOX PANICULATA 'SILVERMINE'

PIERIS 'FLAMING SILVER'

PITTOSPORUM TENUIFOLIUM 'SILVER QUEEN'

PRIMULA AURICULA 'SILVERWAY'

PULMONARIA SACCHARATA 'SILVERADO'

SENECIO CINERARIA 'SILVER DUST'

THYMUS VULGARIS 'SILVER POSIE'

COMMON DESCRIPTIVE TERMS

nivalis, niveus snowy, snow-like
Epimedium × *youngianum* 'Niveum'
Galanthus nivalis
Luzula nivea
Phlox nivalis
Podocarpus nivalis
Rhododendron niveum

---

nobilis noble, renowned
  Laurus nobilis
noctiflorus night-flowering
nodiflorus with flowers at the nodes
  Phyla nodiflora
nodosus with conspicuous nodes
  Geranium nodosum
non-scriptus unmarked
  Hyacinthoides non-scripta (English bluebell)
nootkatensis from the Nootka Sound, British Columbia
  Chamaecyparis nootkatensis
nordmannianus named for Alexander von Nordmann (1803–1866), a German botanist
  Abies nordmanniana
normalis normal (e.g. not variegated, double-flowered etc.)
  Allium paradoxum var. *normale*
northiae named for the British botanical artist and traveller Marianne North (1830–1890)
  Kniphofia northiae

SOME AMERICAN NAMES

alabamensis Alabama
arizonicus Arizona
arkansanus Arkansas
californicus California
carolinianus North Carolina or
  South Carolina
coloradoensis Colorado
floridanus Florida
georgianus Georgia
idahoensis Idaho
illinoinensis Illinois
kentuckiensis Kentucky
ludovicianus Louisiana
marilandicus Maryland
michiganensis Michigan
missouriensis Missouri or the
  Missouri River
nevadensis Nevada or the Sierra
  Nevada (California or Spain)
novae-angliae New England
novi-belgii New York
ohiensis Ohio
oreganus, oregonus Oregon
pensylvanicus Pennsylvania
philadelphicus Philadelphia
tennesseensis Tennessee
texanus, texensis Texas
utahensis Utah
virginianus, virginicus Virginia

novae-zelandiae from New Zealand
  Festuca novae-zelandiae
novi-belgii from New York
  Aster novi-belgii
nucifer nut-bearing
nudicaulis with a bare stem
  Papaver nudicaule
nudiflorus with naked flowers (i.e.
  flowering before the leaves open)
  Jasminum nudiflorum
nudus, nudatus naked
  Nepeta nuda
numidicus from Algeria (the Roman
  province of Numidia)
nummularius like a coin (*nummus*)
  Lysimachia nummularia
nummariifolius with round leaves,
  like a coin
  Olearia nummariifolia

COMMON DESCRIPTIVE TERMS

nutans nodding
Allium nutans
Billbergia nutans
Melica nutans
Ornithogalum nutans

norvegicus Norwegian
notho- false (e.g. *Nothofagus,*
  southern beech*)*
nova-anglica, novae-angliae from
  New England, USA
  Aster novae-angliae
  Eucalyptus nova-anglica

nuttallii named for English botanist,
  Thomas Nuttall (1786–1859)
  Cornus nuttallii
nyctagineus night-flowering
nymansensis from Nymans Garden,
  West Sussex
  Eucryphia × *nymansensis*

O

ob- inverted, reversed
obconicus like an inverted cone
   Primula obconica
obliquus oblique, awry, lopsided
   Nothofagus obliqua
oblongifolius with oblong leaves
   (see below)
   Aster oblongifolius
oblongus, oblongatus oblong
   Cydonia oblonga (quince)
obovatus obovate; inversely ovate
   Paeonia obovata
obscurus obscure, indistinct, dark
   Digitalis obscura
obtusatus blunted
   Hebe obtusata
obtusifolius with blunt leaves
   Cistus × *obtusifolius*
obtusilobus with blunt lobes
   Anemone obtusiloba
obtusus blunt, obtuse
   Chamaecyparis obtusa
obvallaris (literally) walled around
   Narcissus obvallaris
occidentalis western
   Gentiana occidentalis
ocellatus with an eye
   Geranium ocellatum
ochraceus ochre
   Hebe ochracea

ochroleucus yellowish white
   Corydalis ochroleuca
oct-, octo- eight
octopetalus with eight petals
   Dryas octopetala
oculatus with an eye
   Lychnis coronaria Oculata Group
oculus-christi Christ's eye
   Inula oculus-christi
oculus-draconis dragon's eye
   Pinus densiflora 'Oculus-draconis'
ocymoides like *Ocimum* (basil)
   Halimium ocymoides

COMMON DESCRIPTIVE TERMS

odoratus, odorus scented, fragrant
Daphne odora
Galium odoratum (sweet woodruff)
Lathyrus odoratus (sweet pea)
Myrrhis odorata (sweet cicely)
Nymphaea odorata
Reseda odorata (mignonette)
Rosa × *odorata* 'Mutabilis'
Viola odorata (sweet violet)

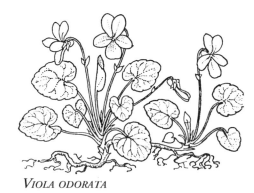

VIOLA ODORATA

WHAT'S IN A NAME?

HABITATS

acraeus living on the heights

agrarius of fields

agrestis rural, rustic; of fields

alpestris growing just below the alpine zone

alpinus alpine, of high mountains

ammophilus sand-loving

amphibius amphibious, growing in water and on land

aquaticus, aquatilis of water; growing by water

arcticus arctic

arenarius of sandy places

arvensis of fields

calcareus chalky, of limestone

calcicola growing on limy soil

campestris, camporum of fields or plains

clivorum of slopes

collinus of hills

cultorum of cultivated land such as gardens

dumetorum of thickets and hedgerows

eremicus of deserts

fluminalis, flumineus, fluviatilis of rivers

fontanus; fontinalis growing in or by springs

frigidus cold; growing in cold locations

glacialis glacial, from cold places

glareosus of gravel

hortensis of gardens

hylaeus of woods

insularis of islands

lacustris of lakes

limosus of marshes

lithophilus stone-loving, growing in stony places

littoralis pertaining to the seashore

maritimus coastal, found near the sea

montanus of mountains

muralis of walls, growing on walls

nemorosus growing in woods

nesophilus island-loving

niphophilus snow-loving

oreophilus mountain-loving

paludosus marshy

palustris of marshes

petraeus growing among rocks

petrophilus rock-loving

pratensis of meadows

riparius of riverbanks

rivalis, rivularis found near brooks

rupestris growing among rocks

rusticanus, rusticus rural, of the country

salinus salty, growing in salty locations

saxatilis, saxosus stony, found among rocks

scopulorum of rocks, cliffs or crags

siliceus growing in sand

spelunca a cave; growing in caves

subalpinus growing below the timber line

sylvaticus, sylvestris of woods

tectorum of roofs

uliginosus marshy, growing in wet places

umbrosus growing in shade

urbanus, urbium urban, of towns

vulcanicus of volcanoes, growing in volcanic soil

CULTIVAR NAMES
Colours: Pink

ASTER ERICOIDES 'PINK CLOUD'

ASTER NOVAE-ANGLIAE 'HARRINGTON'S PINK'

AUBRIETA 'BRESSINGHAM PINK'

CANNA 'PINK SUNBURST'

CLEMATIS 'PINK FLAMINGO'

CLETHRA ALNIFOLIA 'PINK SPIRE'

COLCHICUM 'PINK GOBLET'

CORDYLINE AUSTRALIS 'PINK STRIPE'

DAHLIA 'PINK PASTELLE'

DIANTHUS 'PIKE'S PINK'

ERICA CARNEA 'PINK SPANGLES'

EUCRYPHIA LUCIDA 'PINK CLOUD'

FUCHSIA 'PINK MARSHMALLOW'

GAURA LINDHEIMERI 'SISKIYOU PINK'

GERANIUM SYLVATICUM F. *ROSEUM* 'BAKER'S PINK'

HEBE 'PINK ELEPHANT'

HEMEROCALLIS 'PINK DAMASK'

HYACINTHUS ORIENTALIS 'PINK PEARL'

KOLKWITZIA AMABILIS 'PINK CLOUD'

LAVANDULA ANGUSTIFOLIA 'LODDON PINK'

LILIUM PINK PERFECTION GROUP

MAGNOLIA 'PINKIE'

MONARDA 'CROFTWAY PINK'

PENSTEMON 'HEWELL PINK BEDDER'

PHYGELIUS AEQUALIS 'TREWIDDEN PINK'

POTENTILLA FRUTICOSA 'PINK BEAUTY'

PRUNUS 'PINK PERFECTION'

RIBES SANGUINEUM 'POKY'S PINK'

SORBUS 'PINK-NESS'

COMMON DESCRIPTIVE TERMS

officinalis sold in (apothecaries') shops; used in medicine

Borago officinalis (borage)

Calendula officinalis (pot marigold)

Hyssopus officinalis (hyssop)

Jasminum officinale (jasmine)

Levisticum officinale (lovage)

Melissa officinalis (lemon balm)

Paeonia officinalis (paeony)

Rosmarinus officinalis (rosemary)

Salvia officinalis (sage)

CALENDULA OFFICINALIS

ogisui named for the Japanese botanist Mikinori Ogisu

Epimedium ogisui

ohiensis from Ohio, USA

olbius, olbiensis from the Iles d'Hyères, off the Mediterranean coast of France

Lavatera olbia

oleoides like *Olea* (olive)
    Daphne oleoides
oleifolius with leaves like olive
    Lithodora oleifolia
oleraceus relating to vegetables
    and herbs; of the kitchen garden
    (alternative form of *holeraceus*)
    Brassica oleracea
olig-, oligo- few (see below)
oligophyllus with few leaves
    Ornithogalum oligophyllum
olympicus from Olympus (referring
    to several high mountains: in
    Greece, Asia Minor and America)
    Verbascum olympicum
omeianus, omeiensis from Emei Shan
    (Mount Omei) in Sichuan, China
    Impatiens omeiana
opacus shady, dark
    Ilex opaca
ophiocarpus with snakelike fruits
    Corydalis ophiocarpa
oppositiflorus opposite-flowered
    Gladiolus oppositiflorus
oppositifolius opposite-leaved
    Chrysosplenium oppositifolium
opulifolius with maple-like leaves
    Physocarpus opulifolius
opulus old term for a kind of
    maple tree
    Viburnum opulus
orbiculatus, orbicularis disc-shaped,
    round
    Celastrus orbiculatus

CULTIVAR NAMES
Personal Names
N, O

IRIS 'BROADLEIGH NANCY'
RHODODENDRON 'NAOMI'
CLEMATIS 'NATACHA'
GERANIUM 'NATALIE'
PHLOX MACULATA 'NATASCHA'
ERICA CARNEA 'NATHALIE'
HEBE 'NEIL'S CHOICE'
HEBE 'NICOLA'S BLUSH'
PRIMULA AURICULA 'NIGEL'
CLEMATIS 'OLGA'
DELPHINIUM 'OLIVER'
LILIUM 'OLIVIA'

---

orcadensis from Orkney, Scotland
orchidiflorus with flowers like orchid
    Gladiolus orchidiflorus
orchioides orchid-like
    Iris orchioides

COMMON DESCRIPTIVE TERMS
oreganus, oregonensis, oregonus
from the north-western USA: the
state of Oregon, or the Oregon
division of the Hudson's Bay
Company, which also included
Washington State
Erythronium oregonum
Geranium oreganum
Oxalis oregana
Sedum oreganum
Sedum oregonense

oreophilus mountain-loving
oreoprasum mountain leek
Allium oreoprasum

COMMON DESCRIPTIVE TERMS
orientalis eastern; from the Orient
Crataegus orientalis
Doronicum orientale
Helleborus orientalis
Hyacinthus orientalis
Iris orientalis
Papaver orientale
Pennisetum orientale
Platanus orientalis

---

ornatus showy
Musa ornata
ornithopodus like a bird's foot
Carex ornithopoda 'Variegata'
ortho- straight, upright

orthophyllus with straight leaves
ottawensis from Ottawa, Canada
Berberis × *ottawensis*
ovalifolius with oval leaves
Ligustrum ovalifolium
ovalis oval, broadly elliptic
Carex ovalis

GENUS NAMES
N,O

Nectaroscordum from Greek *nektar*, nectar, and *scorodon*, garlic
Nemophila from Greek *nemos*, wood pasture, and *philo*, to love
Nerine named for a mythical sea-nymph
Nicotiana named for Jean Nicot (1530–1600), a French ambassador in Portugal
Nigella from Latin, a diminutive of *niger*, black
Nymphaea from Greek *nymphaia*, a water-nymph
Olearia Latinized name for Johann Ölschläger (1635–1711), a German horticulturist
Omphalodes from Greek *omphalos*, navel
Onopordum from Greek *onos*, ass, and *porde*, breaking wind
Ophiopogon from Greek *ophis*, snake, and *pogon*, beard
Ornithogalum from Greek *ornis*, bird, and *gala*, milk
Osmanthus from Greek *osme*, fragrance, and *anthos*, flower
Oxalis from Greek *oxys*, acid

Naked ladies *Colchicum*
Nasturtium *Tropaeolum majus*
Navelwort *Omphalodes*
Nettle tree *Celtis*
New England aster *Aster novae-angliae*
New York aster *Aster novi-belgii*
New Zealand flax *Phormium tenax*
Night-scented stock *Matthiola bicornis*
Nightshade *Solanum*
Ninebark *Physocarpus*
Norfolk Island pine *Araucaria
  heterophylla*
Norway maple *Acer platanoides*
Norway spruce *Picea abies*
Oak *Quercus*
Obedient plant *Physostegia virginiana*
Old man's beard *Clematis vitalba*
Oleander *Nerium oleander*
Oleaster *Elaeagnus angustifolia*
Olive *Olea europaea*
Opium poppy *Papaver somniferum*
Orache *Atriplex hortensis*
Orchid *Dactylorhiza, Orchis,
  Ophrys, etc.*
Oregon grape *Mahonia aquifolium*

OPIUM POPPY

Oriental poppy *Papaver orientale*
Orpine *Sedum telephium*
Osier *Salix viminalis*
Ostrich fern *Matteuccia struthiopteris*
Oswego tea *Monarda didyma*
Oxeye daisy *Leucanthemum vulgare*
Oxlip *Primula elatior*

COMMON DESCRIPTIVE TERMS

ovatus ovate, egg-shaped
Asteranthera ovata
Codonopsis ovata
Crassula ovata
Forsythia ovata
Penstemon ovatus

---

ovifer bearing eggs or egg-shaped
  structures (seeds, etc.)

ovinus for sheep
  Festuca ovina

oxonianus, oxoniensis from Oxford
  Geranium × *oxonianum*

oxy- sharp-; sour, acidic (see below)

oxyacanthus with sharp thorns
  Fraxinus angustifolia subsp. *oxycarpa*

oxypetalus with sharp petals

oxysepalus with sharp sepals
  Aquilegia oxysepala

P

pachy- thick

pachycarpus with a thick (thick-shelled) fruit

pachypodus with a thick stalk

pacifica of the Pacific Ocean

   Gypsophila pacifica

padus wild cherry

   Prunus padus

pallescens rather pale, becoming pale

pallidiflorus pale-flowered

   Fritillaria pallidiflora

COMMON DESCRIPTIVE TERMS

pallidus pale

Callistemon pallidus

Hamamelis × *intermedia* 'Pallida'

Iris pallida

Juncus pallidus

Mahonia pallida

palmatus palmate

Acer palmatum

Filipendula palmata

Geranium palmatum

Kirengeshoma palmata

Primula palmatum

Rheum palmatum

---

paludosus marshy

   Epacris paludosa

GENUS NAMES
P

Pachysandra from Greek *pachys*, thick, and *andros*, stamen

Parrotia named for F.W. Parrot (1792–1841), a German naturalist

Parthenocissus from Greek *parthenos*, virgin, and *kissos*, ivy

Paulownia named for Princess Anna Paulowna (1795–1865), daughter of the Czar of Russia

Pelargonium from Greek *pelargos*, a stork

Pennisetum from Latin *penna*, feather, and *seta*, bristle

Penstemon from Greek *pente*, five, and *stemon*, stamen

Perovskia named for V.A. Perovski (1794–1857), a Russian general

Philadelphus from Greek *philadelphos*, brotherly love

Phlox in Greek, a flame

Phormium from Greek *phormion*, mat

Photinia from Greek *phos*, light

Phyllostachys from Greek *phyllon*, leaf, and *stachys*, spike

Physocarpus from Greek *physa*, a bladder, and *karpos*, fruit

Pittosporum from Greek *pitta*, pitch, and *spora*, seed

Platycodon from Greek *platys*, broad, and *kodon*, bell

Polypodium from Greek *polys*, many, and *pous*, foot

Potentilla from Latin *potens*, powerful

Primula from Latin *primus*, first

Pulmonaria from Latin *pulmo*, lung

Pyracantha from Greek *pyr*, fire, and *akantha*, thorn

COMMON DESCRIPTIVE TERMS

palustris of marshes

Calla palustris

Caltha palustris

Epipactis palustris

Euphorbia palustris

Hottonia palustris

Ledum palustre

Pinus palustris

Quercus palustris

pandanifolius with leaves like

  Pandanus (screw pine)

  Eryngium pandanifolium

COMMON DESCRIPTIVE TERMS

paniculatus with flowers arranged
  in panicles

Fuchsia paniculata

Gypsophila paniculata

Hydrangea paniculata

Kolreuteria paniculata

Phlox paniculata

papilio butterfly

  Gladiolus papilio

papyraceus papery

  Narcissus papyraceus

papyrifer paper-bearing

  Betula papyrifera

paradoxus paradoxical; unexpected

  Gentiana paradoxa

paraguayensis from Paraguay

  Azara paraguayensis

CULTIVAR NAMES
World of Music

PELARGONIUM 'ABBA'

CAREX CARYOPHYLLEA 'THE BEATLES'

RHODODENDRON 'BEETHOVEN'

HEMEROCALLIS
'LEONARD BERNSTEIN'

ROSA BENJAMIN BRITTEN
('AUSENCART')

PELARGONIUM 'BOLERO'

RHODODENDRON 'CARMEN'

TULIPA 'CHOPIN'

PRIMULA AURICULA 'CHORISTER'

× *PHYLLIOPSIS* 'COPPELIA'

NARCISSUS 'DÉLIBES'

PHORMIUM 'DUET'

CLEMATIS 'JACQUELINE DU PRÉ'

PAEONIA LACTIFLORA
'SIR EDWARD ELGAR'

PRIMULA AURICULA 'FIGARO'

ROSA 'KATHLEEN FERRIER'

ROSA HANDEL ('MACHA')

PELARGONIUM 'JOSEF HAYDN'

PELARGONIUM 'JAZZ'

CLEMATIS 'KIRI TE KANAWA'

CHAENOMELES 'MADAME BUTTERFLY'

ROSA 'MOZART'

PAPAVER ORIENTALE 'PIZZICATO'

IRIS 'RINGO'

PRUNUS VIRGINIANA 'SCHUBERT'

SAXIFRAGA 'FRANK SINATRA'

TULIPA 'JOHANN STRAUSS'

PRIMULA AURICULA 'SYMPHONY'

PHLOX PANICULATA 'TENOR'

PRIMULA AURICULA 'TOSCA'

CANNA 'VERDI'

pardalianches (literally) poisonous
to panthers; hence leopard's bane
Doronicum pardalianches
pardalinus spotted like a leopard
Lilium pardalinum
parnassicus from Mount Parnassus
Colchicum parnassicum

COMMON DESCRIPTIVE TERMS
parviflorus with small flowers
Aesculus parviflora
Agave parviflora
Cistus parviflorus
Digitalis parviflora
Lithophragma parviflorum
Pinus parviflora

---

parvifolius with small leaves
Eucalyptus parvifolia
patagonicus from Patagonia
Oxalis patagonica
patens spreading
Salvia patens
patulus spreading
Tagetes patula
pauci- few (see below)

COMMON DESCRIPTIVE TERMS
pauciflorus with few flowers
Corylopsis pauciflora
Dierama pauciflorum
Eucalyptus pauciflora
Kniphofia pauciflora
Polemonium pauciflorum

---

paucinervis with few veins
pauciramosus with few branches
Hebe pauciramosa
pauculus very few
pausiacus olive green
pavoninus like a peacock (in colour
– peacock blue – or with an 'eye')
Anemone pavonina
pectinatus like a comb, combed
Euryops pectinatus
pedatisectus cut like a foot
Pinellia pedatisecta
pedatus like a foot (usually meaning
a bird's foot)
Adiantum pedatum
pedemontanus from Piedmont, Italy
Primula pedemontana
peduncularis, pedunculatus with
a peduncle (flower stalk)
Veronica peduncularis
pekinensis from Beijing, China
Phyllostachys aureosulcata f.
pekinensis
pelagicus of the sea
pelargoniiflorus with flowers like
Pelargonium
Erodium pelargoniiflorum
pellucidus translucent
peloponnesiacus from the
Peloponnese region of Greece
Cyclamen peloponnesiacum
peltatus like a *pelta*, a small, half-
moon-shaped shield
Darmera peltata

Paeony, peony *Paeonia*
Pagoda dogwood *Cornus alternifolia*
Palm *Phoenix, Trachycarpus* etc.
Pampas grass *Cortaderia selloana*
Pansy *Viola* × *wittrockiana*
Paper birch *Betula papyrifera*
Paperbark maple *Acer griseum*
Pasque flower *Pulsatilla vulgaris*
Passion flower *Passiflora*
Peach-leaved bellflower *Campanula persicifolia*
Pear *Pyrus communis*
Pearl everlasting *Anaphalis margaritacea*
Pearlbush *Exochorda*
Pebble plants *Lithops*
Pecan *Carya illinoinensis*
Pendulous sedge *Carex pendula*
Père David's maple *Acer davidii*
Periwinkle *Vinca*
Pheasant's eye *Narcissus poeticus* var. *recurvus*
Pickerel weed *Pontaderia cordata*
Piggyback plant *Tolmiea menziesii*
Pimpernel *Anagallis*
Pine *Pinus*

Pineapple broom *Cytisus battandieri*
Pink *Dianthus*
Pitcher plant *Nepenthes; Sarracenia*
Plane *Platanus* (in Scotland: *Acer pseudoplatanus*)
Plantain *Plantago*
Plantain lily *Hosta*
Plume poppy *Macleaya*
Poached-egg flower *Limnanthes douglasii*
Poinsettia *Euphorbia pulcherrima*
Poison ivy *Rhus radicans*
Pokeberry, pokeweed *Phytolacca americana*
Polyanthus *Primula polyantha*
Poplar *Populus*
Poppy *Papaver*
Portugal laurel *Prunus lusitanica*
Pot marigold *Calendula officinalis*
Prickly pear *Opuntia*
Primrose *Primula vulgaris*
Privet *Ligustrum*
Purple loosestrife *Lythrum salicaria*
Purple moor grass *Molinia caerulea*
Purslane *Portulaca*

PINK

WHAT'S IN A NAME?

FLOWER CHARACTERISTICS

acutiflorus with sharply pointed flowers

albiflorus, albiflos white-flowered

amblyanthus with blunt flowers

angustipetalus narrow-petalled

anopetalus with upright petals

apetalus without petals

brevipedunculatus with a short flower-stalk

brevipetalus with short petals

callianthus with beautiful flowers

campaniflorus with bell-shaped flowers

chasmanthus with wide-open flowers

chrysanthus with golden flowers

congestiflorus with closely packed flowers

corymbiflorus with flowers in a corymb

curviflorus with curved flowers

densiflorus densely flowered

dolichostachyus with a long spike

erianthus with woolly flowers

fasciculiflorus with flowers in clusters

flore pleno with double flowers

floridus flowering abundantly

geminiflorus with twin flowers

globiferus bearing globe-shaped clusters

globispicus with globose spikes

glomeratus clustered in a head

grandiflorus large-flowered

lactiflorus with milky-coloured flowers

laetiflorus with bright flowers

latisepalus with broad sepals

laxiflorus loose-flowered

longiflorus with long flowers

longipetalus with long petals

lophanthus with crested flowers

macranthus with large flowers

macropetalus with large petals

megalanthus with large flowers

micranthus with small flowers

nudiflorus with naked flowers (i.e. flowering before the leaves open)

oppositiflorus opposite-flowered

oxypetalus with sharp petals

pallidiflorus pale-flowered

paniculatus with flowers arranged in panicles

parviflorus with small flowers

platyanthus with broad flowers

platypetalus with broad (or flat) petals

pleniflorus with double (full) flowers

plenus double (full)

pogonanthus with bearded flowers

polyandrus with many anthers

polyanthus with many flowers

psilostemon with bare or smooth stamens

racemosus with flowers in racemes

radiatus with rays; radiating in form

rariflorus with scattered flowers

recurvus, recurvatus recurved, curved backwards

reflexus bent back

sessiliflorus with sessile flowers

stenanthera with narrow anthers

stylosus with a conspicuous or large style

thyrsiflorus with flowers in a thyrse

thyrsoides like a thyrse

umbellatus umbelliferous

viridiflorus with green flowers

COMMON DESCRIPTIVE TERMS

pendulus pendulous, hanging
Betula pendula (silver birch)
Carex pendula (pendulous sedge)
Cercidiphyllum japonicum f. *pendulum*
Dierama pendulum
Fraxinus excelsior 'Pendula'
Pyrus salicifolia 'Pendula'

---

penicillaris with a paintbrush-like
  tuft of hair
  Melica penicillaris
pensylvanicus from Pennsylvania,
  USA
  Acer pensylvanicum
penta- five (see below)
pentapetalus with five petals
  Geum pentapetalum
pentaphyllus with five leaves
  or leaflets
  Acer pentaphyllum
perdicarius of partridges
  Oxalis perdicaria
peregrinans wandering abroad
  Libertia peregrinans
peregrinus foreign, exotic
  Tropaeolum peregrinum

COMMON DESCRIPTIVE TERMS

perennis perennial
Bellis perennis
Linum perenne
Oenothera perennis
Pterocephalus perennis

CULTIVAR NAMES
Personal Names
P

DIGITALIS PURPUREA 'PAM'S CHOICE'
CLEMATIS 'PAMELA'
PRIMULA 'OUR PAT'
GERANIUM 'PATRICIA'
PAPAVER ORIENTALE 'PATTY'S PLUM'
CRATAEGUS LAEVIGATA
  'PAUL'S SCARLET'
HEBE 'PAULA'
CLEMATIS 'PAULINE'
FUCHSIA 'PEGGY'
ROSA 'PENELOPE'
PELARGONIUM 'PENNY'
DAHLIA 'PETER'
GENTIANA ASCLEPIADEA 'PHYLLIS'
FUCHSIA 'PEGGY'
HYDRANGEA MACROPHYLLA 'PIA'
ROSA 'POLLY'

COMMON DESCRIPTIVE TERMS

perfoliatus perfoliate (with the
  leaf surrounding the stem)
Claytonia perfoliata
Eupatorium perfoliatum
Parahebe perfoliata
Smyrnium perfoliatum

---

perforatus pierced with holes or pores
  Hypericum perforatum
periclymenum honeysuckle (Greek)
  Lonicera periclymenum
permixtus much mixed

P

PRIMULA AURICULA
'PADDLIN MADELEINE'

PERSICARIA VIRGINIANA
'PAINTER'S PALETTE'

PRIMULA AURICULA 'PALEFACE'

RHODODENDRON 'PANDA'

HEMEROCALLIS 'PARDON ME'

MALUS DOMESTICA (APPLE) 'PEAR APPLE'

MALUS DOMESTICA
'PEASGOOD'S NONSUCH'

NARCISSUS 'PEEPING TOM'

MAGNOLIA 'PEPPERMINT STICK'

AUCUBA JAPONICA 'PEPPERPOT'

RHODODENDRON 'PERSIL'

PHLOX BIFIDA 'PETTICOAT'

MALUS DOMESTICA 'PIG'S SNOUT'

HOSTA 'PINEAPPLE UPSIDE
DOWN CAKE'

PHLOX × *ARENDSII* 'PING PONG'

LOBELIA 'PINK ELEPHANT'

FUCHSIA 'PINK GOON'

PHORMIUM 'PINK PANTHER'

PELARGONIUM 'PINK PARADOX'

IRIS ENSATA 'PIN STRIPE'

HELENIUM PIPSQUEAK ('BLOPIP')

IRIS 'POGO'

RHODODENDRON 'POLAR BEAR'

PINUS THUNBERGII 'PORKY'

ORIGANUM VULGARE 'POLYPHANT'

PHLOX PANICULATA 'POPEYE'

IRIS 'PUMPIN' IRON'

PENSTEMON 'PURPLE PASSION'

SAXIFRAGA 'PURPLE PIGGY'

---

perralderianus named for a French naturalist, Henri de la Perraudière (1831–1861)
  Epimedium perralderianum
persicifolius with leaves like *Prunus persica* (peach tree)
  Campanula persicifolia

COMMON DESCRIPTIVE TERMS
persicus from Iran (Persia)
Cyclamen persicum
Fritillaria persica
Morina persica
Parrotia persica

---

persimilis very similar (e.g. to another species)
  Crataegus persimilis 'Prunifolia'
perspicuus transparent
perulatus with conspicuous bud-scales (perules)
  Enkianthus perulatus
peruvianus Peruvian
  Scilla peruviana (named by Linnaeus in error – it actually comes from the Mediterranean

---

COMMON DESCRIPTIVE TERMS
petiolaris with a (long) leaf stalk (petiole)
Helichrysum petiolare
Hydrangea anomala subsp. *petiolaris*
Primula petiolaris
Tilia 'Petiolaris'

petraeus growing among rocks
   Quercus petraea
petrophilus rock-loving
   Epacris petrophila
phaeus dusky
   Geranium phaeum
phello- cork
   Euonymus phellomanus
philadelphicus from Philadelphia,
   USA
   Erigeron philadelphicus
philippinensis from the Philippines
   Lilium philippinensis
phillyreoides like *Phillyrea*
   Quercus phillyreoides
phlogopappus with seeds like *Phlox*
   Olearia phlogopappa
phoeniceus Phoenician; purple-red
   (from the Phoenician purple dye
   made from the mollusc *Murex*)
   Verbascum phoeniceum
phu rotten-smelling; acrid
   Valeriana phu 'Aurea'
phyllocephalus with a leafy head
   Carex phyllocephala
physalodes like *Physalis*
   Nicandra physalodes
picturatus embellished; variegated
   Crocus vernus 'Picturatus'
pictus (literally) painted; vividly
   coloured
   Acer pictum
pileatus with a cap
   Lonicera pileata

VALERIANA PHU

COMMON DESCRIPTIVE TERMS

pilosus with long soft hairs, shaggy
   Dipsacus pilosus
   Genista pilosa
   Luzula pilosa
   Papaver pilosum

pilosissimus very shaggy
   Heuchera pilosissima
pilulifer bearing little balls
   (small globular fruits)
   Carex pilulifera
pimeleoides like the Australian
   evergreen shrub *Pimelea*
   Hebe pimeleoides

pimpinellifolius with pinnate leaves like the umbelliferous plant *Pimpinella*
    Rosa pimpinellifolia
pinaster wild pine
    Pinus pinaster
pinguifolius with fat leaves
    Hebe pinguifolia
pinifolius with leaves like pine
pinnatifidus pinnately divided
    Quercus dentata 'Pinnatifida'
pinnatisectus pinnately cut
    Erigeron pinnatisectus

PINNATIFIDUS

COMMON DESCRIPTIVE TERMS
pinnatus pinnate
Epimedium pinnatum
Lavandula pinnata
Mahonia pinnata
Santolina pinnata

---

piperitus like pepper
    Mentha × *piperita*
pisifer (literally) bearing peas
    Chamaecyparis pisifera
planifolius with flat leaves
    Iris planifolia
planipes with a flat stalk or foot
    Euonymus planipes
planiscapus with a flat or even scape (leafless stem)
    Ophiopogon planiscapus
plantagineus like *Plantago* (plantain)
    Hosta plantaginea

platanifolius with leaves like *Platanus* (plane tree)
    Ranunculus platanifolius
planus flat
    Eryngium planum
platanoides like *Platanus*
    Acer platanoides
platy- broad, flat
platyanthus with broad flowers
    Geranium platyanthum
platyglossus with a broad tongue
    Phyllostachys platyglossa
platypetalus with broad (or flat) petals
    Epimedium platypetalum

COMMON DESCRIPTIVE TERMS
platyphyllus, platyphyllos with broad (or flat) leaves
Galanthus platyphyllus
Lilium auratum var. *platyphyllum*
Limonium platyphyllum
Tilia platyphyllos

---

pleniflorus with double (full) flowers
    Kerria japonica 'Pleniflora'

plenus double (flowers); full (see also page 81, *flore pleno*)
 Clematis 'Purpurea Plena Elegans'
plicatus folded or pleated
 Viburnum plicatum
plinianus named for the 1st-century AD Roman scholar Pliny the Elder
 Arundo pliniana
plumarius feathery
 Dianthus plumarius
plumbaginoides like *Plumbago*
 Ceratostigma plumbaginoides

COMMON DESCRIPTIVE TERMS
plumosus feathered
 Gymnocarpium dryopteris 'Plumosum'
 Muscari comosum 'Plumosum'
 Sambucus racemosa 'Plumosa Aurea'

pluricaulis with several stems
 Sedum pluricaule
poculiformis cup-shaped
 Galanthus nivalis Poculiformis Group

CULTIVAR NAMES
Gardeners and Plantsmen

THYMUS PULEGIOIDES 'BERTRAM ANDERSON'

HELLEBORUS × HYBRIDUS 'HELEN BALLARD'

CROCUS CHRYSANTHUS 'E.A. BOWLES'

GERANIUM MACULATUM 'BETH CHATTO'

HEUCHERA 'HELEN DILLON'

FUCHSIA CHARLIE DIMMOCK ('FONCHA')

FORSYTHIA 'BEATRIX FARRAND'

ARTEMISIA LUDOVICIANA 'VALERIE FINNIS'

PENSTEMON 'MARGERY FISH'

MAHONIA × MEDIA 'LIONEL FORTESCUE'

PENSTEMON 'GEOFF HAMILTON'

OENOTHERA 'PENELOPE HOBHOUSE'

ACER PALMATUM 'COLLINGWOOD INGRAM'

VINCA MINOR F. ALBA 'GERTRUDE JEKYLL'

ROSA 'LAWRENCE JOHNSTON'

COLCHICUM AUTUMNALE 'NANCY LINDSAY'

GAZANIA 'CHRISTOPHER LLOYD'

NARCISSUS 'CEDRIC MORRIS'

PULMONARIA 'LEWIS PALMER'

MECONOPSIS CAMBRICA 'FRANCES PERRY'

ARTEMISIA ARBORESCENS 'FAITH RAVEN'

VIOLA 'VITA'

DIANTHUS 'LADY SALISBURY'

BERGENIA 'ERIC SMITH'

FUCHSIA 'GEOFFREY SMITH'

PRIMULA AURICULA 'ANNE SWITHINBANK'

ROSA GRAHAM THOMAS ('AUSMAS')

ASTER NOVI-BELGII 'PERCY THROWER'

DIANTHUS 'ALAN TITCHMARSH'

VIOLA 'TONY VENISON'

GERANIUM × OXONIANUM 'ROSEMARY VEREY'

SCABIOSA CAUCASICA 'MISS WILLMOTT'

podagricus swollen at the base (literally 'gouty')
  Jatropha podagrica
podophyllus with stalked leaves
  Rodgersia podophylla
poeticus, poetarum relating to (usually Classical) poets
  Narcissus poeticus
pogonanthus with bearded flowers
  Geranium pogonanthum
polifolius grey-leaved
  Andromeda polifolia
poluninianus named for Oleg Polunin (1914–1985), the British botanist, author and plant collector
  Saxifraga poluniniana
poly- many (see below)
polyacanthus with many thorns
  Opuntia polyacantha
polyandrus with many anthers
  Phytolacca polyandra
polyanthus with many flowers
  Jasminum polyanthum
polycarpus with many fruits
polychromus many-coloured
  Euphorbia polychroma
polymorphus with many forms
  Persicaria polymorpha
polyneurus with many veins
  Primula polyneura
polyphyllus with many leaves
  Paris polyphylla
polytrichus with many hairs
  Draba polytricha

pomifer bearing apples
ponderosus heavy
  Pinus ponderosa

COMMON DESCRIPTIVE TERMS

ponticus from Pontus, an area south of the Black Sea, in Asia Minor
Artemisia pontica
Daphne pontica
Fritillaria pontica
Quercus pontica
Rhododendron ponticum

---

populifolius with leaves like *Populus* (poplar)
  Cistus populifolius
porphyrophyllus with purple leaves
  Sambucus nigra f. *porphyrophylla*
populneus of poplars
  Hoheria populnea
portenschlagianus named for Franz von Portenschlag-Ledermayer (1777–1827), an Austrian botanist
  Campanula portenschlagiana
poscharskyanus named for Gustav Poscharsky (1832–1914), a German gardener
  Campanula poscharskyana
potaninii named for Grigori Potanin (1835–1920), a Russian naturalist and explorer
  Indigofera potaninii
potentilloides like *Potentilla*
  Geranium potentilloides

CULTIVAR NAMES
Foreign Expressions
P

PAPILLON (*French*) BUTTERFLY
(*IRIS SIBIRICA* 'PAPILLON')

PARELMOER (*Dutch*)
MOTHER OF PEARL
(*GAULTHERIA MUCRONATA* 'PARELMOER')

PERLE D'AZUR (*French*) AZURE PEARL
(*CLEMATIS* 'PERLE D'AZUR')

PERLMUTTERSCHALE (*German*)
MOTHER OF PEARL
(*PLATYCODON GRANDIFLORUS*
'PERLMUTTERSCHALE')

PETIT FAUCON (*French*)
LITTLE FALCON
(*CLEMATIS* PETIT FAUCON
('EVISIX'))

PLUIE DE FEU (*French*) RAIN OF FIRE
(*HEUCHERA* 'PLUIE DE FEU')

POLARSOMMER (*German*)
ARCTIC SUMMER
(*VERBASCUM BOMBYCIFERUM*
'POLARSOMMER')

PORZELLAN (*German*)
PORCELAIN
(*NEPETA* 'PORZELLAN')

PRÄRIENACHT (*German*)
PRAIRIE NIGHT
(*MONARDA* 'PRÄRIENACHT')

PREZIOSA (*Italian*) PRECIOUS
(*HYDRANGEA* 'PREZIOSA')

PRIMAVERA (*Italian*) SPRING
(*CAMELLIA JAPONICA* 'PRIMAVERA')

PRINZ HEINRICH (*German*)
PRINCE HENRY
(*ANEMONE* × *HUPEHENSIS* VAR. *JAPONICA*
'PRINZ HEINRICH')

PÜNKTCHEN (*German*)
LITTLE DOT
(*MISCANTHUS SINENSIS* 'PÜNKTCHEN')

PURPURLANZE (*German*)
PURPLE LANCE
(*ASTILBE CHINENSIS* VAR. *TAQUETII*
'PURPURLANZE')

COMMON DESCRIPTIVE TERMS

praecox precocious, developing early
Chimonanthus praecox
Clematis 'Praecox'
Cytisus × *praecox*
Hydrangea paniculata 'Praecox'
Prunus incisa 'Praecox'
Rhododendron 'Praecox'
Stachyurus praecox

praemorsus as if bitten off
praestans distinguished, excellent
   Tulipa praestans
praeteritus past
praetermissus neglected, overlooked,
   omitted
   Dactylorhiza praetermissa
praetextus fringed, bordered
prasinus green like a leek

NURSERIES AND NURSERYMEN

backhouseanus James Backhouse (1794–1869), English nurseryman

beesianus Bees, nursery and seed supplier in Cheshire

burkwoodii Arthur (1888–1951) and Albert Burkwood, brothers who ran a nursery at Kingston upon Thames

durandii Durand Frères, French nursery

fraseri John Fraser (1750–1811), London nurseryman and plant collector in North America

frikartii Carl Ludwig Frikart (1879-1964), Swiss nurseryman

jackmanii George Jackman (1801–1869) and his son, British clematis breeders in Woking, Surrey

lawsonianus Charles Lawson (1794–1873), Scottish nurseryman who raised Lawson's cypress from seed

lemoinei Victor Lemoine (1823–1911) and his son Emile (1862–1942), French nurserymen

lizei Lizé Frères, nursery in Nantes, France

soulangeanus Etienne Soulange-Bodin (1774–1846), French horticulturist

sprengeri Karl Sprenger (1846–1917), German nurseryman working in Italy

standishii John Standish (1814–1875), British nurseryman

vanhouttei Louis Van Houtte (1810–1876), Belgian nurseryman,

veitchii, veitchianus Veitch & Sons, 19th-century nurseries in Exeter and London

vilmorinii Vilmorin-Andrieux, French nursery and seed supplier

COMMON DESCRIPTIVE TERMS

pratensis of meadows

Cardamine pratensis (lady's smock)

Geranium pratense (meadow cranesbill)

Lathyrus pratensis

Pulsatilla pratensis

primulifolius with leaves like *Primula*

Campanula primulifolia

primuloides like *Primula*

Mimulus primuloides

princeps princely, most distinguished

Artemisia princeps

prismaticus prism-shaped

Iris prismatica

proboscideus with a proboscis or snout

Arisarum proboscideum

procerus high, tall

Ulmus procera (English elm)

COMMON DESCRIPTIVE TERMS

procumbens prostrate

Asarina procumbens

Fuchsia procumbens

Gaultheria procumbens

Juniperus procumbens

procurrens spreading, running

Arabis procurrens

proliferus free-flowering; or reproducing by offshoots

Primula prolifera

prolificus prolific, fruitful
  Hypericum prolificum
propinquus related, neighbouring
  Phyllostachys propinqua
prostratus prostrate
  Veronica prostrata
proteiflorus with flowers like *Protea*
  Eryngium proteiflorum
prunifolia with leaves like *Prunus*
  Crataegus persimilis 'Prunifolia'
przewalskii named for a Russian
  explorer and naturalist, Nikolai
  Przewalski (1839–1888)
  Ligularia przewalskii
pseud-, pseudo- false (see below)
pseudacorus like ('false') *Acorus*
  (sweet flag)
  Iris pseudacorus
pseudoacacia like ('false') *Acacia*
  Robinia pseudoacacia
pseudocamellia like ('false') *Camellia*
  Stewartia pseudocamellia
pseudocyperus like ('false') *Cyperus*
  Carex pseudocyperus
pseudodictamnus like ('false')
  Dictamnus
  Ballota pseudodictamnus
pseudonarcissus like ('false') *Narcissus*
  Narcissus pseudonarcissus
pseudoplatanus like ('false') *Platanus*
  (plane)
  Acer pseudoplatanus (sycamore)
psilostemon with smooth stamens
  Geranium psilostemon

psittacinus like a parrot
  Alstroemeria psittacina
ptarmicus, ptarmicoides sneeze-
  inducing
  Achillea ptarmica
  Aster ptarmicoides
puberulus slightly downy
  Helenium puberulum

COMMON DESCRIPTIVE TERMS
pubescens, pubiger downy, with
  soft hair
Betula pubescens
Epimedium pubigerum
Primula × *pubescens*
Syringa pubescens

---

pudicus shy
  Mimosa pudica

COMMON DESCRIPTIVE TERMS
pulcher, pulchellus beautiful, pretty
Coleonema pulchrum
Correa pulchella
Crocus pulchellus
Deutzia pulchra
Dodecathon pulchellum
Gaillardia pulchella
Geranium pulchrum
Heuchera pulchella

---

pulcherrimus very beautiful
  Dierama pulcherrimum
pulchra see *pulcher*

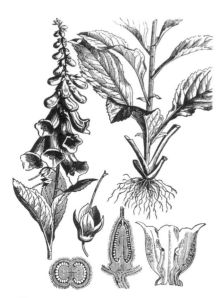

DIGITALIS PURPUREA

pulegius, pulegioides derived
from Latin *pulex* (flea),
indicating flea-repellent qualities
in aromatic plants
  Mentha pulegium (Pennyroyal)
  Thymus pulegioides
pullus dark in colour
  Campanula pulla

COMMON DESCRIPTIVE TERMS
pulverulentus dusty
Cistus × *pulverulentus*
Eucalyptus pulverulenta
Primula pulverulenta
Sambucus nigra 'Pulverulenta'

---

pulvinaris, pulvinatus like a cushion
  Myosotis pulvinaris

COMMON DESCRIPTIVE TERMS
pumilus dwarf
Astilbe chinensis var. *pumila*
Cortaderia selloana 'Pumila'
Ficus pumila
Iris pumila
Lilium pumilum
Pinus pumila

punctatus spotted
Anthemis punctata subsp. *cupaniana*
Campanula punctata
Lysimachia punctata
Phyllostachys nigra var. *punctata*

---

pungens piercing, pungent, ending
  in a hard sharp point
  Elaeagnus pungens
puniceus purplish crimson
  Clianthus puniceus

COMMON DESCRIPTIVE TERMS
purpurascens purplish, becoming
  purple; **purpuratus** clad in purple;
  purpureus purple
Bergenia purpurascens
Clematis 'Purpurea Plena Elegans'
Corylus maxima 'Purpurea'
Digitalis purpurea
Echinacea purpurea
Eupatorium purpureum
Foeniculum vulgare 'Purpureum'
Osmanthus heterophyllus 'Purpureus'
Weigela florida 'Foliis Purpureis'

purpusii named for German plant-collecting brothers Carl (1853–1941) and Josef (1860–1932) Purpus
  Lonicera × *purpusii*
pustulatus blistered or pimply
  Begonia pustulata 'Argentea'

COMMON DESCRIPTIVE TERMS
pygmaeus pygmy, dwarf, very small
Borago pygmaea
Chamaecyparis obtusa 'Pygmaea'
Lewisia pygmaea
Nymphaea 'Pygmaea Helvola'
Penstemon hirsutus var. *pygmaeus*

---

pyramidalis pyramid-shaped
  Campanula pyramidalis

COMMON DESCRIPTIVE TERMS
pyrenaicus, pyrenaeus from the Pyrenees (on the borders of France and Spain)
Fritillaria pyrenaica
Geranium pyrenaicum
Hepatica nobilis var. *pyrenaica*
Horminum pyrenaicum
Lilium pyrenaicum
Ornithogalum pyrenaicum
Quercus pyrenaica

---

pyrethrifolius with leaves like
  Pyrethrum (now *Tanacetum*)
  Leptinella pyrethrifolia

CULTIVAR NAMES
Some American States
CANNA 'ALASKA'
PELARGONIUM 'ARIZONA'
FREMONTODENDRON 'CALIFORNIA GLORY'
HEDERA HELIX 'CAROLINA CRINKLE'
NYMPHAEA 'COLORADO'
LILIUM 'CONNECTICUT KING'
PRIMULA AURICULA 'DAKOTA'
CANNA 'DELAWARE'
RHODODENDRON 'FLORIDA'
CUPHEA LLAVEA 'GEORGIA SCARLET'
CAMELLIA JAPONICA 'HAWAII'
PSEUDOTSUGA MENZIESII 'IDAHO GEM'
MORUS 'ILLINOIS EVERBEARING'
NYMPHAEA 'INDIANA'
PAEONIA LACTIFLORA 'KANSAS'
IRIS 'KENTUCKY BLUEGRASS'
PHLOX DIVARICATA 'LOUISIANA PURPLE'
MAGNOLIA 'MARYLAND'
ARCTOSTAPHYLOS UVA-URSI 'MASSACHUSETTS'
PHILADELPHUS 'MINNESOTA SNOWFLAKE'
HEMEROCALLIS 'MISSOURI BEAUTY'
ROSA 'NEVADA'
GERANIUM SANGUINEUM 'NEW HAMPSHIRE PURPLE'
IRIS 'OREGON SKIES'
FUCHSIA 'TENNESSEE WALTZ'
FUCHSIA 'TEXAS LONGHORN'
PINUS BANKSIANA 'WISCONSIN'
CANNA 'WYOMING'

Q

quadrangularis, quadrangulatus with
four angles
Chimonobambusa quadrangularis
Fraxinus quadrangulata
quadratus square; in fours; in four
Restio quadratus
quadridentatus with four teeth
quadrifolius with four leaves
or leaflets
Paris quadrifolia
quadripinnatus four times pinnate
Lophosoria quadripinnata
quamash from a Native American
word meaning 'sweet'
Camassia quamash
quarciticus like quartz
quercifolius with leaves like
Quercus (oak)
Hydrangea quercifolia
quinatus in fives
Akebia quinata
quinquefolius with five leaves
Parthenocissus quinquefolia
quitoensis from Quito, Ecuador
Solanum quitoense
quinquelobatus with five lobes
Pelargonium quinquelobatum
quintuplinervius with five veins
Meconopsis quintuplinervia
quotidianus everyday, common

CULTIVAR NAMES
Nurseries and Plant Breeders

HELLEBORUS × *HYBRIDUS* ASHWOOD
GARDEN HYBRIDS

ROSA PAT AUSTIN ('AUSMUM')'

LOBELIA 'BEES' FLAME'

DIASCIA 'BLACKTHORN APRICOT'

DICENTRA 'ADRIAN BLOOM'

IRIS 'BROADLEIGH ROSE'

GERANIUM 'SUE CRÛG'

CALAMAGROSTIS × *ACUTIFLORA*
'KARL FOERSTER'

PULMONARIA SACCHARATA
'GLEBE COTTAGE BLUE'

DAPHNE 'VALERIE HILLIER'

POTENTILLA FRUTICOSA
'HOPLEYS ORANGE'

GERANIUM MACRORRHIZUM
'INGWERSEN'S VARIETY'

MACLEAYA MICROCARPA
'KELWAY'S CORAL PLUME'

MALUS × *PURPUREA* 'LEMOINEI'

VIBURNUM OPULUS
'NOTCUTT'S VARIETY'

PRUNUS DOMESTICA (PLUM)
'RIVERS'S EARLY PROLIFIC'

SAXIFRAGA 'RIVERSLEA'

ESCALLONIA 'SLIEVE DONARD'

ASTRANTIA MAJOR
'SUNNINGDALE VARIEGATED'

DIGITALIS PURPUREA
'SUTTON'S APRICOT'

CLEMATIS 'JOHN TREASURE'

ECHINOPS RITRO 'VEITCH'S BLUE'

SPIRAEA JAPONICA
'ANTHONY WATERER'

ROSA 'HARRY WHEATCROFT'

Alpine strawberry *Fragaria vesca*
Apple *Malus domestica*
Apricot *Prunus armeniaca*
Banana *Musa*
Bilberry *Vaccinium myrtillus*
Blackberry *Rubus fruticosus*
Blackcurrant *Ribes nigrum*
Blueberry *Vaccinium corymbosum*
Cape gooseberry *Physalis peruviana*
Cherry *Prunus avium*
Chinese gooseberry (kiwi fruit)
   Actinidia deliciosa
Cranberry *Vaccinium macrocarpon*
Damson *Prunus insititia* varieties
Date *Phoenix dactylifera*
Fig *Ficus carica*
Gooseberry *Ribes uva-crispa*
Grape *Vitis vinifera*
Grapefruit *Citrus × paradisi*
Greengage *Prunus domestica*
   varieties
Guava *Psidium guajava*
Kiwi fruit *Actinidia deliciosa*
Lemon *Citrus limon*
Lime *Citrus aurantifolia*
Loganberry *Rubus loganobaccus*
Loquat *Eriobotrya japonica*
Mango *Mangifera indica*
Medlar *Mespilus germanica*
Melon *Cucumis melo*
Mulberry, black *Morus nigra*
Nectarine *Prunus persica* var.
   nectarina
Orange *Citrus sinensis* (see also
   Seville orange)
Papaya (paw-paw) *Carica papaya*
Passion fruit *Passiflora edulis*
Peach *Prunus persica*

Pear *Pyrus communis*
Persimmon *Diospyros virginiana*
Physalis *Physalis peruviana*
Pineapple *Ananas comosus*
Plum *Prunus domestica*
Pomegranate *Punica granatum*
Prickly pear *Opuntia*

PEAR

Quince *Cydonia oblonga*
Raspberry *Rubus idaeus*
Redcurrant *Ribes rubrum*
Rhubarb *Rheum × hybridum*
Satsuma *Citrus unshui*
Seville orange *Citrus aurantium*
Sloe *Prunus spinosa*
Strawberry *Fragaria × ananassa*
Watermelon *Citrullus lanatus*

R

COMMON DESCRIPTIVE TERMS

racemosus with flowers in racemes
Aralia racemosa
Cornus racemosa
Danae racemosa
Exochorda racemosa
Nepeta racemosa
Passiflora racemosa
Sambucus racemosa

---

raddeanus named for Gustav Radde
    (1831–1903), a German
    naturalist
    Campanula raddeana
radiatus with rays; radiating
radicans with rooting stems
    Campsis radicans
radula a scraper or file
    Pelargonium 'Radula'
ramiflorus flowering on branches
ramosissimus much branched
    Tamarix ramosissima

COMMON DESCRIPTIVE TERMS

ramosus branched
Anthericum ramosum
Asphodelus ramosus
Francoa ramosa
Sasaella ramosa

ramulosus twiggy
    Olearia ramulosa
ranunculoides like *Ranunculus*
    Anemone ranunculoides
raphanifolius with leaves like
    Raphanus (radish)
    Cardamine raphanifolia
rapunculoides like a little turnip
    Campanula rapunculoides
rariflorus with scattered flowers
reclinatus bent back or down
    Phoenix reclinata
rectifolius with erect leaves
rectus upright
    Clematis recta
recurvus, recurvatus recurved,
    curved backwards
    Narcissus poeticus var. *recurvus*
redivivus revived
    Lunaria rediviva
reducta reduced
    Sorbus reducta

COMMON DESCRIPTIVE TERMS

reflexus bent back
Correa reflexa
Elaeagnus × *reflexa*
Geranium reflexum
Lindera reflexa

---

refractus bent back
refulgens shining
regalis royal; fit for a king
    Lilium regale

Quaking-grass *Briza*
Quamash *Camassia*
Queen Anne's lace *Anthriscus sylvestris*
Quince *Cydonia oblonga*
Ragged robin *Lychnis flos-cuculi*
Ragwort *Senecio*
Ramsons *Allium ursinum*
Rangoon creeper *Quisqualis indica*
Redbud *Cercis canadensis*
Red campion *Silene dioica*
Red-hot poker *Kniphofia*
Red valerian *Centranthus ruber*
Reed *Phragmites*
Reedmace *Typha*
Restharrow *Ononis*
Ribbon grass *Phalaris*
Rock cress *Arabis*
Rock jasmine *Androsace*
Rock purslane *Calandrinia umbellata*
Rock samphire *Crithmum maritimum*
Rock-rose *Cistus, Helianthemum*
Rocky Mountain juniper *Juniperus scopulorum*
Rose *Rosa*
Rose of Sharon *Hypericum*
Rosebay willowherb *Chamerion angustifolium*
Rosemary *Rosmarinus officinalis*

ROWAN

Royal fern *Osmunda regalis*
Rowan *Sorbus aucuparia*
Rubber plant *Ficus elastica*
Rue *Ruta graveolens*
Rush *Juncus*
Russian olive *Elaeagnus*
Russian sage *Perovskia atriplicifolia*
Russian vine *Fallopia baldschuanica*
Rye *Elymus*

reginae of the queen; *reginae olgae* named for Queen Olga of Greece (1851–1926)
  Galanthus reginae-olgae
  Strelitzia reginae
regius royal
  Juglans regia
regularis regular, uniform

rehderi, rehderianus named for Alfred Rehder (1863–1949), a dendrologist at the Arnold Arboretum in Massachusetts
  Clematis rehderiana
remotus remote, scattered
reniformis kidney-shaped
  Pelargonium reniforme

repandens, repandus with wavy
  or turned-up edges
  Cyclamen repandum
  Taxus baccata 'Repandens'
replicatus folded back

Common descriptive terms
repens, reptans creeping
Ajuga reptans
Coprosma repens
Gypsophila repens
Mazus reptans
Melinis repens
Ononis reptans
Polemonium reptans

---

requienii named for Esprit Requien
  (1788–1851), a French naturalist
  Mentha requienii (Corsican mint)
resinifer, resinosus resinous
reticulatus netted
  Iris reticulata
retinodes tenacious, retaining
  Acacia retinodes
retroflexus bent back
retusus with a rounded or blunt tip
  Daphne tangutica Retusa Group
revolutus rolled or turned back
  Erythronium revolutum
rex king
  Begonia rex
rhaeticus from the Rhaetian
  Alps (Switzerland/Austria)
  Geum × *rhaeticum*
rhamnoides like *Rhamnus*
  (buckthorn)
  Hippophae rhamnoides
rhizomatosus rhizomatous
  Epimedium rhizomatosum
rhodopensis from the Rhodope
  Mountains in Bulgaria
  Haberlea rhodopensis
rhombeus, rhombicus, rhombiformis
  rhomboidal, diamond-shaped
  Hedera rhombea
rhombifolius with diamond-shaped
  leaves
  Cissus rhombifolia
rhytidophyllus with wrinkled leaves
  Viburnum rhytidophyllum

COMMON DESCRIPTIVE TERMS

rigens, rigidus rigid, stiff; **rigescens,
rigidulus** rather stiff
Diascia rigescens
Euphorbia rigida
Gazania rigens
Hebe rigidula
Leontodon rigens
Polystichum rigens
Verbena rigida

---

rigidifolius with stiff leaves
rimosus cracked
ringens open-mouthed, gaping,
wide open
Mimulus ringens
riparius of riverbanks
Carex riparia
ritualis ritual, of ceremonies
Kniphofia ritualis

rivalis, rivularis growing near brooks
Cirsium rivulare 'Atropurpureum'
Geum rivale (water avens)
riversleaianum named for Prichard's
Riverslea nursery at Christchurch,
Dorset
Geranium × *riversleaianum*
rivinianus after Augustus Rivinus,
the Latinized name of
A. Bachmann (1652–1722),
a German botanist
Viola riviniana (common
dog violet)
robbiae named for Mary Anne
Robb (1829–1912); she is
reputed to have brought *Euphorbia
amygdaloides* var. *robbiae* back from
Turkey in her hat-box: hence its
nickname, Mrs Robb's bonnet
robinsonianus (see panel below)

WILLIAM ROBINSON (1838–1935)

William Robinson began his career as a
gardener in Ireland before moving to England,
where he became one of the most influential
plantsmen of his day. A passionate advocate
of naturalistic gardening, he founded the
periodical *The Garden*, and his books *The
Wild Garden* (1870) and *The English Flower
Garden* (1883) remain classics. Plant names
associated with him include *Anemone nemorosa*
'Robinsoniana'. *Clematis* 'Gravetye Beauty'
and *Geranium himalayense* 'Gravetye' are
named for his home, Gravetye Manor, in West
Sussex. His head gardener there was clematis
expert Ernest Markham.

Quisqualis from Latin *quis?* (what?), and *qualis?* (what kind?)

Ranunculus from a Latin diminutive of *rana*, a frog

Reseda from Latin *resedare*, to heal

Rhaphiolepis from Greek *rhaphis*, needle, and *lepis*, scale

Rhodochiton from Greek *rhodo*, red, and *chiton*, a tunic or cloak

Rhododendron from Greek *rhodon*, rose, and *dendron*, tree

Ribes from Arabic *ribas*, acid

Robinia named for Jean Robin (1550–1629), French royal gardener

Rodgersia named for John Rodgers (1812–1882), American naval officer

Romneya named for Thomas Romney Robinson (1792–1882), Irish astronomer

Rosmarinus from Latin *ros*, dew, and *marinus*, maritime

Rudbeckia named for Olof Rudbeck (1630–1702), Swedish botanist

---

robur oakwood, hardwood; strength
  Quercus robur

---

robustus robust, strong

Eremurus robustus

Fargesia robusta

Grevillea robusta

Malus × *robusta*

Taxus cuspidata 'Robusta'

---

romanus Roman

rooperi named for Edward Rooper (1818–1854), a British soldier serving in South Africa
  Kniphofia rooperi

rosa-sinensis Chinese rose
  Hibiscus rosa-sinensis

---

roseus, rosea like a rose, rose-coloured

Alcea rosea

Centranthus ruber 'Roseus'

Deutzia × *rosea*

Erodium × *variabile* 'Roseum'

Primula rosea

Rhodiola rosea

Wisteria floribunda 'Rosea'

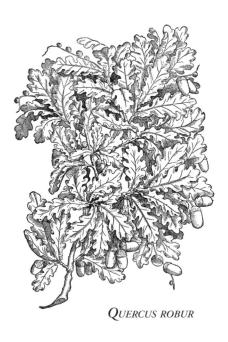

QUERCUS ROBUR

CULTIVAR NAMES
Foreign Expressions
R

RAKET (*Dutch*) ROCKET
(*POPULUS ALBA* 'RAKET')

RAUBRITTER (*German*)
ROBBER KNIGHT
(*ROSA* 'RAUBRITTER')

REGENBOGEN (*German*) RAINBOW
(*HELIANTHEMUM* 'REGENBOGEN')

REINE DU JOUR (*French*)
QUEEN OF THE DAY
(*PHLOX MACULATA* 'REINE DU JOUR')

REINE ROUGE (*French*) RED QUEEN
(*MENTHA* × *PIPERITA* 'REINE ROUGE')

RIJNSTROOM (*Dutch*) RIVER RHINE
(*GEUM* 'RIJNSTROOM')

RÖDE KLOKKE (*Flemish*) RED BELL
(*PULSATILLA VULGARIS* 'RÖDE KLOKKE')

ROODKAPJE (*Dutch*) RED CAP
(*DIANTHUS* 'ROODKAPJE')

ROSA ERFÜLLUNG (*German*)
PINK ZENITH
(*ASTER AMELLUS* 'ROSA ERFÜLLUNG')

ROSAKÖNIGIN (*German*)
PINK QUEEN
(*HELIANTHEMUM* 'ROSAKÖNIGIN')

ROSENKUPPEL (*German*) ROSY DOME
(*ORIGANUM* 'ROSENKUPPEL')

ROSENPOKAL (*German*) ROSY
GOBLET
(*ARTEMISIA* 'ROSENPOKAL')

ROSENROT (*German*) ROSE RED
(*PERSICARIA CAMPANULATA* 'ROSENROT')

ROSENSCHLEIER (*German*)
ROSY VEIL
(*ARTEMISIA* 'ROSENSCHLEIER')

ROTBLUM (*German*) RED FLOWER
(*BERGENIA* 'ROTBLUM')

ROTFUCHS (*German*) RED FOX
(*VERONICA SPICATA* 'ROTFUCHS')

ROTGOLD (*German*) RED GOLD
(*HELENIUM* 'ROTGOLD')

ROTKÄPPCHEN (*German*)
LITTLE RED RIDING HOOD
(*PRUNELLA GRANDIFLORA*
'ROTKÄPPCHEN')

ROTKEHLCHEN (*German*) ROBIN
(*HYDRANGEA MACROPHYLLA*
'ROTKEHLCHEN')

ROTKOPF (*German*) RED HEAD
(*SEMPERVIVUM* 'ROTKOPF')

ROTKUGEL (*German*) RED BALL
(*ORIGANUM* 'ROTKUGEL')

ROTSILBER (*German*) RED SILVER
(*MISCANTHUS SINENSIS* 'ROTSILBER')

ROTSTIEL (*German*) RED STEM
(*ATHYRIUM FILIX-FEMINA* 'ROTSTIEL')

ROTSTRAHLBUSCH (*German*)
RED SUNBEAM BUSH
(*PANICUM VIRGATUM*
'ROTSTRAHLBUSCH')

RUBINZWERG (*German*)
RUBY DWARF
(*HELENIUM* 'RUBINZWERG')

rosmarinifolius with leaves like
 Rosmarinus (rosemary)
 Ozothamnus rosmarinifolius
rostratus beaked
rosularis with rosettes
 Plantago major 'Rosularis'
rotatus shaped like a wheel

Common descriptive terms

rotundifolius with round leaves
Bellis rotundifolia
Campanula rotundifolia
Lathyrus rotundifolius
Ligustrum japonicum 'Rotundifolium'
Origanum rotundifolium
Pellaea rotundifolia
Prostanthera rotundifolia
Tithonia rotundifolia

rotundus round
rubellus light red

Common descriptive terms

ruber red
Acer rubrum
Alnus rubra
Astrantia major 'Rubra'
Atriplex hortensis var. *rubra*
Centranthus ruber
Epimedium × *rubrum*
Escallonia rubra
Filipendula rubra
Fritillaria imperialis 'Rubra'
Imperata cylindrica 'Rubra'
Pulmonaria rubra
Pulsatilla rubra
Quercus rubra

rubescens reddish, reddening
 Acer rubescens
rubiginosus rusty
 Rosa rubiginosa

rubra see *ruber*

rubrifolius with red leaves
　Plantago major 'Rubrifolia'

rubromarginatus edged with red
　Phyllostachys rubromarginata

rufinervis with red veins
　Acer rufinerve

rufus, rufescens reddish
　Fargesia rufa

rugosus, rugulosus wrinkled
　Rosa rugosa

runcinatus with teeth pointing
　towards the base
　Persicaria runcinata

COMMON DESCRIPTIVE TERMS

rupestris rock-loving
Erodium rupestre
Leptospermum rupestre
Potentilla rupestris
Sedum rupestre

---

rupicola rock-dweller
　Penstemon rupicola

rupifragus rock-breaking
　Papaver rupifragum

ruscifolius with leaves like *Ruscus*
　(butcher's broom)
　Coriaria ruscifolia

russelianus named for John Russell
　(1766–1839), 6th Duke of Bedford
　Phlomis russeliana

russicus from Russia
　Echium russicum

WHAT'S IN A NAME?

COLOUR: YELLOW AND GOLD

auratus, aureatus decorated with gold
aureopictus (literally) gold-painted
aureus gold, golden
canarinus canary yellow
cerinus waxy yellow
chrys-, chryso- gold
chrysographes marked with gold
citrinus lemon yellow
croceus saffron yellow
flavens yellow
flavescens, flavicans, flavidus yellowish
flavissimus very yellow
flavovirens yellowish green
flavus pure, pale yellow
fulvus, fulvellus, fulvescens, fulvidus
　tawny yellow, yellowish brown
helvolus greyish yellow
icterinus jaundice yellow
isabellinus tawny yellow
luridus dirty yellow
luteolus, lutescens pale yellow
luteus yellow
ochraceus brownish yellow
stramineus like straw, straw-coloured
sulphureus sulphur yellow
vitellinus egg-yolk yellow
xanth-, xantho- yellow
xanthinus yellow

---

rusticanus, rusticus rural
　Armoracia rusticana
Magnolia × *soulangeana* 'Rustica
　Rubra'
ruthenicus from Ruthenia, a region
　of Russia (and used more generally
　to mean Russian)
　Echinops ritro subsp. *ruthenicus*

S

sabatius from Savona, on the
Ligurian coast of north-west Italy
Convolvulus sabatius

saccatus like a bag; with sacs
Plectranthus saccatus

saccharatus, saccharinus sweet, sugary
Acer saccharinum
Pulmonaria saccharata

saccharum sugar cane
Acer saccharum

sachalinensis from the island of
Sakhalin, off the east coast of Russia
Lilium sachalinense

sagittalis, sagittatus arrow-shaped
Genista sagittalis

sagittifolius with arrow-shaped leaves
Hedera hibernica 'Sagittifolia'

salicarius like *Salix* (willow)
Lythrum salicaria (purple
loosestrife)

COMMON DESCRIPTIVE TERMS

salicifolius with leaves like willow
Pyrus salicifolia 'Pendula'
Aucuba japonica f. *longifolia*
'Salicifolia'
Buphthalmum salicifolium
Cotoneaster salicifolius
Hebe salicifolia
Magnolia salicifolia

CORNUS SANGUINEA

salignus like willow
Callistemon salignus

salinus salty, growing in salty places

salsillus salted

saluenensis from the region of
Nu Jiang (the Salween river) in
Yunnan, China, and Burma
Nomocharis saluenensis

salviifolius with leaves like *Salvia*
Cistus salviifolius

sambucinus like *Sambucus* (elder)

samius from the Greek island
of Samos
Phlomis samia

sancti-johannis St John
Anthemis sancti-johannis

sanctus holy

COMMON DESCRIPTIVE TERMS

sanguineus bloody, blood-red
Cornus sanguinea
Geranium sanguineum
Heuchera sanguinea
Ribes sanguineum

---

sapidus tasty, savoury
sarajevensis from Sarajevo
  Knautia sarajevensis
sarcocaulis with fleshy stems
  Crassula sarcocaulis
sardiniensis from the Mediterranean
  island of Sardinia

COMMON DESCRIPTIVE TERMS

sargentii, sargentianus named for
  the American dendrologist Charles
  Sprague Sargent (1841–1927), first
  director of the Arnold Arboretum
  in Massachusetts, USA
Hydrangea aspera subsp. *sargentiana*
Lilium sargentiae
Magnolia sargentiana var. *robusta*
Malus toringo subsp. *sargentii*
Prunus sargentii
Rhododendron sargentianum
Sorbus sargentiana

---

sarmaticus from Sarmatia (an old
  name for a region covering part
  of eastern Poland, Belarus and
  the Ukraine)
  Campanula sarmatica

sarmentosus with runners
  Androsace sarmentosa
sarniensis from Guernsey,
  Channel Islands
  Nerine sarniensis

CULTIVAR NAMES
Personal Names
S

ROSA SALLY'S ROSE ('CANREM')
LIRIOPE MUSCARI 'SAMANTHA'
KNIPHOFIA 'SAMUEL'S SENSATION'
HAMAMELIS VERNALIS 'SANDRA'
STREPTOCARPUS 'SARAH'
HEMEROCALLIS 'SEBASTIAN'
PRIMULA AURICULA 'SHEILA'
TULIPA 'SHIRLEY'
FUCHSIA 'SIOBHAN'
POLEMONIUM 'SONIA'S BLUEBELL'
FUCHSIA 'SOPHIE'S SILVER LINING'
SEQUOIADENDRON GIGANTEUM
'LITTLE STAN'
FUCHSIA 'SOPHIE LOUISE'
VACCINIUM CORYMBOSUM 'STANLEY'
CAMPANULA POSCHARSKYANA 'STELLA'
ERODIUM 'STEPHANIE'
HEDYCHIUM DENSIFLORUM 'STEPHEN'
IRIS SIBIRICA 'STEVE'
PRIMULA AURICULA 'SUSANNAH'
LONICERA PERICLYMENUM 'SWEET SUE'
MAGNOLIA 'SUSAN'
PRIMULA 'SUNSHINE SUSIE'
CAMPANULA CARPATICA 'SUZIE'
NARCISSUS 'SUZY'
IRIS 'BROADLEIGH SYBIL'
CAMELLIA JAPONICA 'SYLVIA'

CULTIVAR NAMES
Colours: White

BERGENIA 'BRESSINGHAM WHITE'

BUDDLEJA DAVIDII
'WHITE PROFUSION'

CALLUNA VULGARIS 'WHITE LAWN'

CAMPANULA LACTIFLORA
'WHITE POUFFE'

CLEMATIS 'WHITE MOTH'

CORTADERIA SELLOANA
'WHITE FEATHER'

DIANTHUS 'HAYTOR WHITE'

ECHINACEA PURPUREA 'WHITE SWAN'

ERICA CARNEA 'SPRINGWOOD WHITE'

ERYTHRONIUM CALIFORNICUM
'WHITE BEAUTY'

GERANIUM CLARKEI
'KASHMIR WHITE'

GYPSOPHILA FASTIGIATA
'WHITE FESTIVAL'

HYDRANGEA MACROPHYLLA
'LANARTH WHITE'

IRIS 'WHITE CITY'

LAMIUM MACULATUM 'WHITE NANCY'

LATHYRUS LATIFOLIUS 'WHITE PEARL'

LILIUM LONGIFLORUM
'WHITE AMERICAN'

NARCISSUS 'WHITE LION'

PAEONIA LACTIFLORA 'WHITE WINGS'

PELARGONIUM 'WHITE BOAR'

PENSTEMON 'WHITE BEDDER'

PHLOX PANICULATA
'WHITE ADMIRAL'

PIERIS JAPONICA 'WHITE RIM'

RHODODENDRON 'LODER'S WHITE'

SORBUS 'WHITE WAX'

VIBURNUM TINUS 'FRENCH WHITE'

COMMON DESCRIPTIVE TERMS

sativus sown, planted, cultivated
Castanea sativa (sweet chestnut)
Coriandrum sativum (coriander)
Crocus sativus (saffron crocus)
Lathyrus sativus
Medicago sativa (alfalfa, lucerne)

---

saxatilis found among rocks
   Primula saxatilis
saxifragus stone-breaking
   Petrorhagia saxifraga
saximontanus from the Rocky
   Mountains, North America
   Aquilegia saximontana
saxosus stony, growing among stones
   Gentiana saxosa
scaber, scabrosus rough, gritty
   Eccremocarpus scaber
scabiosifolius with leaves like *Scabiosa*
   (scabious)
   Patrinia scabiosifolia
scabriusculus slightly rough or gritty
   Fuchsia scabriuscula
scalaris like a ladder
   Sorbus scalaris

COMMON DESCRIPTIVE TERMS

scandens climbing
Celastrus scandens
Cobaea scandens
Dicentra scandens
Hibbertia scandens
Senecio scandens

scandicus, scandinavicus from
Scandinavia
Primula scandinavica
scariosus shrivelled
sceleratus harmful
schillingii named for the British
plantsman Tony Schilling
Euphorbia schillingii
schizo- split, divided
schoenoprasum (literally) rush-like
leek
Allium schoenoprasum (chives)
schubertii named for Gotthilf von
Schubert (1780–1860), a German
naturalist and plant collector
Allium schubertii
scilloides like *Scilla*
Puschkinia scilloides var. *libanotica*
scilloniensis from the Isles of Scilly,
England
Olearia × *scilloniensis*
scirpoides like *Scirpus* (club rush)
Equisetum scirpoides
sclero- hard
scolopendrius millipede
Asplenium scolopendrium
scoparius a broom or sweeper
Cytisus scoparius
scopulinus little rock
Erigeron scopulinus
scopulorum of rocks, cliffs or crags
Juniperus scopulorum
scorodonius of garlic
Teucrium scorodonia

WHAT'S IN A NAME?

PLANT HUNTERS AND COLLECTORS

armandii as *davidii*, below
davidii, davidianum Abbé Jean Pierre
Armand David (1826–1900)
douglasii, douglasianus David
Douglas (1799–1834)
drummondii James (1786–1863) and
Thomas (1793–1835) Drummond
endressii Philip Endress (1806–1831)
fargesii Paul Farges (1844–1912)
farreri Reginald Farrer (1880–1920)
forrestii George Forrest (1873–1932)
fortunei Robert Fortune (1812–
1880)
giraldianus, giraldii Giuseppe Giraldi
(1848–1901)
haastii Sir Johann von Haast
(1824–1887)
jeffreyi John Jeffrey (1826–1854)
lancasteri Roy Lancaster
lindheimeri Ferdinand Lindheimer
(1802–1879)
lyallii David Lyall (1817–1895)
mariesii Charles Maries (1851–
1902)
ogisui Mikinori Ogisu
purpusii Carl (1853–1941) and Josef
(1860–1932) Purpus
sieboldianus, sieboldii Philipp von
Siebold (1796–1866)
turczaninowii Nicolai Turczaninov
(1796–1864)
wardii Frank Kingdon Ward (1885–
1958)
whittallii Edward Whittall (1851-
1917)
wilsonianus, wilsonii E.H. ('Chinese')
Wilson (1876–1930)

Saffron crocus *Crocus sativus*
Sagebrush *Artemisia*
St John's wort *Hypericum*
Sallow *Salix caprea*
Sandwort *Arenaria*
Scarborough lily *Cyrtanthus elatus*
Scotch thistle *Onopordum acanthium*
Scots pine *Pinus sylvestris*
Sea buckthorn *Hippophae rhamnoides*
Sea holly *Eryngium*
Sea lavender *Limonium platyphyllum*
Sea pink *Armeria maritima*
Sedge *Carex*
Shasta daisy *Leucanthemum × superbum*
Shepherd's purse *Capsella bursa-pastoris*
Shoo-fly plant *Nicandra physalodes*
Shuttlecock fern *Matteuccia struthiopteris*
Silver birch *Betula pendula*
Skullcap *Scutellaria*
Smoke bush *Cotinus coggygria*
Snakebark maple *Acer capillipes; Acer davidii*
Snakeshead fritillary *Fritillaria meleagris*
Snapdragon *Antirrhinum*
Snowball tree *Viburnum opulus* 'Roseum'
Snowberry *Symphoricarpos*
Snowdrop *Galanthus*
Snow-in-summer *Cerastium tomentosum*

Snowy mespilus *Amelanchier*
Soft shield fern *Polystichum setiferum*
Solomon's seal *Polygonatum multiflorum*
Southernwood *Artemisia abrotanum*
Spanish broom *Spartium junceum*
Speedwell *Veronica*
Spider flower *Cleome*
Spider plant *Chlorophytum comosum*
Spindle *Euonymus europaeus*
Spleenwort *Asplenium*
Spotted laurel *Aucuba japonica*
Spring snowflake *Leucojum vernum*
Spruce *Picea*
Spurge *Euphorbia*
Spurge laurel *Daphne laureola*
Squill *Scilla*
Star-of-Bethlehem *Ornithogalum umbellatum*
Statice *Limonium*
Stock *Matthiola*
Stonecrop *Sedum*
Strawberry tree *Arbutus*
Sugar maple *Acer saccharum*
Sunflower *Helianthus annuus*
Swan River daisy *Brachyscome*
Sweet alyssum *Lobularia maritima*
Sweet briar *Rosa rubiginosa*
Sweet cicely *Myrrhis odorata*
Sweet pea *Lathyrus odoratus*
Sweet rocket *Hesperis matronalis*
Sweet sultan *Amberboa*
Sweet William *Dianthus barbatus*
Sweet woodruff *Galium odoratum*

SHEPHERD'S PURSE

scorpioides like a scorpion
  Myosotis scorpioides
scoticus Scottish
  Daboecia cantabrica subsp. *scotica*
scutellaroides like *Scutellaria*
  Solenostemon scutellaroides (coleus)
sectus cut
  Carex secta
secunda secund (arranged on one side only)
  Echeveria secunda var. *glauca*
sedifolius with leaves like *Sedum*
  Aster sedifolius
seguieri, seguierianus named for J.F. Séguier (1703–1784), a French botanist
  Euphorbia seguieriana
selloanus named for Friedrich Sello (1789–1831), a German naturalist and plant collector
  Cortaderia selloana
semiplenus semi-double
  Camassia leichtlinii 'Semiplena'
semperflorens always flowering
  Begonia semperflorens

COMMON DESCRIPTIVE TERMS

sempervirens evergreen
Buxus sempervirens
Cupressus sempervirens
Helictotrichon sempervirens
Iberis sempervirens
Lonicera sempervirens
Sequoia sempervirens

CULTIVAR NAMES
Anniversaries

CLEMATIS ANNIVERSARY ('PYNOT')
IRIS SIBIRICA 'ANNIVERSARY'
DAHLIA 'ANNIVERSARY BALL'
IRIS 'ANNIVERSARY CELEBRATION'
FUCHSIA 'DIAMOND CELEBRATION'
FUCHSIA 'GOLDEN ANNIVERSARY'
HOSTA 'GOLDEN ANNIVERSARY'
LAMIUM MACULATUM GOLDEN ANNIVERSARY ('DELLAM')
ROSA 'GOLDEN ANNIVERSARY'
ROSA GOLDEN CELEBRATION ('AUSGOLD')
PELARGONIUM 'GOLDEN WEDDING'
RHODODENDRON 'GOLDEN WEDDING'
ROSA HAPPY ANNIVERSARY ('BEDFRANC')
ROSA PEARL ANNIVERSARY ('WHITSTON')
ROSA RUBY ANNIVERSARY ('HARBONNY')
ROSA RUBY CELEBRATION ('PEAWINNER')
ASTRANTIA 'RUBY WEDDING'
CAMELLIA × *WILLIAMSII* 'RUBY WEDDING'
FUCHSIA 'RUBY WEDDING'
LEPTOSPERMUM SCOPARIUM 'RUBY WEDDING'
ROSA 'RUBY WEDDING'
BUDDLEJA 'SILVER ANNIVERSARY'
CAMELLIA JAPONICA 'SILVER ANNIVERSARY'
ROSA SILVER ANNIVERSARY ('POULARI')
ROSA 'SILVER WEDDING'

NARCISSUS 'SATSUMA'

MONARDA 'SCORPION'

PINUS SYLVESTRIS 'SCRUBBY'

MALUS DOMESTICA (APPLE) 'SCRUMPTIOUS'

PINUS STROBUS 'SEA URCHIN'

PLATYCODON GRANDIFLORUS 'SENTIMENTAL BLUE'

IRIS 'SERENGETI SPAGHETTI'

ROSA SEXY REXY ('MACREXY')

ASTRANTIA MAJOR SUBSP. *INVOLUCRATA* 'SHAGGY'

PAEONIA 'SHAGGY DOG'

GERANIUM MACULATUM 'SHAMEFACE'

IRIS 'SHAMPOO'

MALUS DOMESTICA 'SHEEP'S NOSE'

GERANIUM SANGUINEUM 'SHEPHERD'S WARNING'

JUNIPERUS SCOPULORUM 'SKYROCKET'

PELARGONIUM 'SMALL FORTUNE'

NARCISSUS 'SMALL TALK'

PRIMULA AURICULA 'THE SNEEP'

DAHLIA 'SNEEZY'

PRIMULA AURICULA 'SNOOTY FOX'

FUCHSIA 'SNOW WHITE' (AND 'BASHFUL', 'DOC', 'DOPEY', 'GRUMPY', 'HAPPY', 'SLEEPY' AND 'SNEEZY'!)

DIANTHUS BARBATUS 'SOOTY'

DIANTHUS 'SOPS-IN-WINE'

PENSTEMON 'SOUR GRAPES'

PHLOX PANICULATA 'SPEED LIMIT'

HEMEROCALLIS 'SPIDERMAN'

JUNIPERUS COMMUNIS 'SPOTTY SPREADER'

IRIS 'STAIRWAY TO HEAVEN'

SANGUISORBA TENUIFOLIA 'STAND UP COMEDIAN'

IBERIS SEMPERVIRENS 'STARKERS'

MAGNOLIA 'STAR WARS'

OMPHALODES CAPPADOCICA 'STARRY EYES'

PRIMULA AURICULA 'STORMIN NORMAN'

HEMEROCALLIS 'STRAWBERRY FIELDS FOREVER'

PLANTAGO LANCEOLATA 'STREAKER'

HOSTA 'STRIPTEASE'

MALUS DOMESTICA 'STUB NOSE'

SAXIFRAGA 'SUGAR PLUM FAIRY'

PAPAVER ORIENTALE 'SULTANA'

LEUCANTHEMUM × *SUPERBUM* 'SUNNY SIDE UP'

IRIS 'SUPERSTITION'

PHORMIUM 'SURFER'

QUERCUS PALUSTRIS 'SWAMP PYGMY'

---

sempervivoides like *Sempervivum* (houseleek)

senescens growing old
  Allium senescens

sensibilis, sensitivus sensitive
  Onoclea sensibilis

senticosus thorny

septem- seven (see below)

septemfidus divided into seven
  Gentiana septemfida

septemlobus with seven lobes
  Kalopanax septemlobus

septentrionalis northern (from the seven stars of the Plough constellation)
   Betula albosinensis var. *septentrionalis*
serbicus from Serbia
   Ranunculus serbicus
sericeus silky
   Daphne sericea
sericifer silk-bearing

COMMON DESCRIPTIVE TERMS

serotinus late
Dianthus serotinus
Lonicera periclymenum 'Serotina'
Populus × *canadensis* 'Serotina'
Prunus serotina

serratus serrated, saw-toothed; **serratifolius** with serrated leaves; **serrula** a small saw; **serrulatus** with small saw-like teeth
Hydrangea serrata
Penstemon serrulatus
Photinia serratifolia
Prunus serrula
Zelkova serrata

---

sessilis sessile, stalkless
   Trillium sessile
sessiliflorus with sessile flowers
   Libertia sessiliflora
sessilifolius with sessile leaves
   Lobelia sessilifolia

setaceus bristly
   Asparagus setaceus
setchuenensis from Sichuan, China
   Deutzia setchuenensis
setifer bearing bristles
   Polystichum setiferum
setosus bristly
   Iris setosa
setulosus with small bristles
   Crassula setulosa
sexstylosus with six styles
   Hoheria sexstylosa

COMMON DESCRIPTIVE TERMS

sibericus, sibiricus from Siberia
Claytonia sibirica
Cornus alba 'Sibirica'
Iris sibirica
Ligularia sibirica
Nepeta sibirica
Scilla siberica

---

sibthorpianus, sibthorpii named for the English botanist Humphrey Sibthorp (1713–1797) and his son John (1758–1796)
   Fritillaria sibthorpiana
   Primula vulgaris subsp. *sibthorpii*
siculum Sicilian, from Sicily
   Nectaroscordum siculum
sieberi named for the Prague naturalist Franz Sieber (1789–1844)
   Crocus sieberi

GENUS NAMES
S

Santolina from Latin *sanctum linum*, holy flax

Saponaria from Latin *sapo*, soap

Sarcococca from Greek *sarco-*, flesh, and *kokkos*, berry

Saxifraga from Latin *saxum*, rock, and *frangere*, to break

Schizanthus from Greek *schizo*, to divide, and *anthos*, flower

Schizostylis from Greek *schizo*, to divide, and *stylis*, style

Scutellaria from Latin *scutella*, a small dish

Sedum from Latin *sedere*, to sit

Sempervivum from Latin *semper*, always, and *vivus*, alive

Senecio from Latin *senex*, old man

Shibataea named for Keita Shibata (1877–1949), Japanese botanist

Skimmia from the Japanese name *Shikimi*

Stachyurus from Greek *stachys*, spike, and *oura*, tail

Stewartia named for John Stuart (1713–1792), 3rd Earl of Bute and British Prime Minister. Unfortunately Linnaeus misspelled his name

Stokesia named for Dr Jonathan Stokes (1755–1831), English botanical author

Streptocarpus from Greek *streptos*, twisted, and *karpos*, fruit

Symphoricarpos from Greek *symphorein*, bear together, and *karpos*, fruit

Syringa from Greek *syrinx*, pipe

sieboldianus, sieboldii named for Philipp von Siebold (1796–1866), a German doctor who had a special interest in the plants of Japan and introduced many into Europe

Acer sieboldianum

Clematis florida var. *sieboldiana*

Dryopteris sieboldii

Hosta sieboldiana

Magnolia sieboldii

Primula sieboldii

Viburnum sieboldii

sikkimensis from Sikkim in the eastern Himalayas

Allium sikkimensis

Euphorbia sikkimensis

Malus sikkimensis

Musa sikkimensis

Primula sikkimensis

sikikianus from the Japanese island of Shikoku

siliceus growing in sand

silvaticus, silvestris of woods, growing wild (see also *sylvaticus*)

silvicola an inhabitant of woods

similis similar

Lonicera similis var. *delavayi*

simplex simple, unbranched

Actaea simplex

simplicicaulis with unbranched stems

Centaurea simplicicaulis

CULTIVAR NAMES
Foreign Expressions
S

SANGO-KAKU (*Japanese*) CORAL SHELL
(*ACER PALMATUM* 'SANGO-KAKU')

SATSUKI (*Japanese*) AZALEA
(*PELARGONIUM* 'SATSUKI')

SCHARLACHGLUT (*German*)
SCARLET GLOW
(*ROSA* 'SCHARLACHGLUT ')

SCHNEEHAUBE (*German*) SNOW CAP
(*ARABIS ALPINA* SUBSP.
CAUCASICA 'SCHNEEHAUBE')

SCHNEEWITTCHEN (*German*)
SNOW WHITE
(*ERIGERON* 'SCHNEEWITTCHEN')

SCHWEFELBLÜTE (*German*)
FLOWERS OF SULPHUR
(*ACHILLEA* 'SCHWEFELBLÜTE')

SEEIGEL (*German*) SEA URCHIN
(*FESTUCA GLAUCA* 'SEEIGEL')

SEMPRE AVANTI (*Italian*)
ALWAYS IN FRONT
(*NARCISSUS* 'SEMPRE AVANTI')

SHIROFUJI (*Japanese*)
MOUNT FUJI WHITE
(*HYDRANGEA SERRATA* 'SHIROFUJI')

SHIROTAE (*Japanese*) SNOW WHITE
(*PRUNUS* 'SHIROTAE')

SILBERFEDER (*German*)
SILVER FEATHER
(*MISCANTHUS SINENSIS* 'SILBERFEDER')

SILBERLICHT (*German*) SILVER LIGHT
(*BERGENIA* 'SILBERLICHT')

SILBERLOCKE (*German*) SILVER CURL
(*ABIES KOREANA* 'SILBERLOCKE')

SILBERSCHMELZE (*German*)
MOLTEN SILVER
(*ERICA* × *DARLEYENSIS*
'SILBERSCHMELZE')

SINEE DOZHD (*Russian*) BLUE RAIN
(*CLEMATIS* 'SINEE DOZHD')

SOLEIL D'OR (*French*) GOLDEN SUN
(*PYRACANTHA* 'SOLEIL D'OR')

SOMMERSCHNEE (*German*)
SUMMER SNOW
(*ANAPHALIS TRIPLINERVIS*
'SOMMERSCHNEE')

SONNENKIND (*German*) SUN CHILD
(*COREOPSIS* 'SONNENKIND')

SONNENWENDE (*German*) SOLSTICE
(*OENOTHERA FRUTICOSA* SUBSP. *GLAUCA*
'SONNENWENDE')

SOUVENIR DE... (*French*)
MEMORY OF...
(*ROSA* 'SOUVENIR DU DOCTEUR
JAMAIN')

SPITZENTÄNZERIN (*German*)
BALLERINA
(*HELIOPSIS HELIANTHOIDES* VAR. *SCABRA*
'SPITZENTÄNZERIN')

STELLA DE ORO (*Italian*) GOLD STAR
(*HEMEROCALLIS* 'STELLA DE ORO')

STRAUSSENFEDER (*German*)
OSTRICH FEATHER
(*ASTILBE* 'STRAUSSENFEDER')

simplicifolius with simple (i.e. not divided or lobed) leaves

Meconopsis simplicifolia

simplicissimus completely undivided or unbranched

CULTIVAR NAMES
Weddings

ALSTROEMERIA 'BLUSHING BRIDE'

LATHYRUS LATIFOLIUS 'BLUSHING BRIDE'

LAVATERA × *CLEMENTII* 'BLUSHING BRIDE'

GERANIUM DALMATICUM 'BRIDAL BOUQUET'

NARCISSUS 'BRIDAL CROWN'

IRIS SIBIRICA 'BRIDAL JIG'

DIANTHUS 'BRIDAL VEIL'

FUCHSIA 'BRIDESMAID'

PELARGONIUM 'BRIDESMAID'

RHODODENDRON 'BRIDESMAID'

FUCHSIA 'HAPPY WEDDING DAY'

HYDRANGEA PANICULATA 'OCTOBER BRIDE'

CLEMATIS 'THE BRIDE'

EXOCHORDA × *MACRANTHA* 'THE BRIDE'

GAURA LINDHEIMERI 'THE BRIDE'

GERANIUM VERSICOLOR 'THE BRIDE'

GLADIOLUS 'THE BRIDE'

PRUNUS INCISA 'THE BRIDE'

NARCISSUS 'TROUSSEAU'

CAMPANULA PUNCTATA 'WEDDING BELLS'

CHRYSANTHEMUM 'WEDDING DAY'

ROSA 'WEDDING DAY'

COMMON DESCRIPTIVE TERMS

sinensis, sinicus Chinese

Buxus sinica var. *insularis*

Citrus sinensis (orange tree)

Corylopsis sinensis

Geranium sinense

Magnolia sieboldii subsp. *sinensis*

Miscanthus sinensis

Nyssa sinensis

Wisteria sinensis

---

sinuatus, sinuosus wavy-edged

Limonium sinuatum

sino- Chinese

sitchensis from Sitka, Alaska

Picea sitchensis

smilacinus like *Smilax*

Disporum smilacinum

sobolifer with creeping stems that form roots

Jovibarba sobolifera ('hen and chickens' houseleek)

socialis growing in colonies

Ledebouria socialis

solandri named for Daniel Solander (1736–1782), a biologist who travelled with Captain Cook on the *Endeavour*

Olearia solandri

solaris sun-loving

soldanellus like a small coin

Carex soldanella

solidus solid, dense

Corydalis solida

somnifer sleep-bringing
   Papaver somniferum
sonchifolius with leaves like *Sonchus*
   (sow-thistle)
   Francoa sonchifolia
songaricus from Dzungaria in north-
   west China, bordering Kazakhstan
   Clematis songarica
soulangeanus named for Etienne
   Soulange-Bodin (1774–1846),
   a French horticulturist
   Magnolia × *soulangeana*
spathaceus with a spathe
   Salvia spathacea
spathulatus spoon-shaped
   Moraea spathulata
spathulifolius with spoon-shaped
   leaves
   Sedum spathulifolium

COMMON DESCRIPTIVE TERMS

speciosus showy, splendid
Chaenomeles speciosa
Colchicum speciosum
Crocus speciosus
Lilium speciosum
Ribes speciosum
Tropaeolum speciosum

spectabilis spectacular
Dicentra spectabilis
Lampranthus spectabilis
Primula spectabilis
Sedum spectabile

WHAT'S IN A NAME?

COLOUR: BLUE AND PURPLE

amethystinus violet-coloured
atropurpureus dark purple
azureus sky blue
betinus purple like beetroot
caeruleus, coeruleus bright, deep
   blue
caesius light greyish blue, lavender
   blue
cobaltinus cobalt blue
coelestinus sky blue
cyaneus, cyanus clear, bright blue
ianthinus violet-blue
indigoticus deep, dark blue
lilacinus lilac-coloured
lividus leaden blue-grey
ostrinus rich purple
pavoninus peacock blue
porphyro-, porphyreus purple
purpurascens purplish, becoming
   purple
purpuratus clad in purple
purpureus purple
tyrius purple
violaceus violet

spectatus beheld; esteemed
spelunca cave
   Microlepia speluncae
-spermus -seeded
sphaericus spherical
sphaerocarpus with globose fruits
sphaerocephalus with globose heads
   Allium sphaerocephalon
sphegodes like a wasp
   Ophrys sphegodes
sphen- wedge-shaped

Common descriptive terms

spicatus spicate, bearing a spike
Hedychium spicatum
Liatris spicata
Liriope spicata
Mentha spicata (spearmint)
Veronica spicata

spicifer spike-bearing
spiciformis spike-shaped
spiculatus with fine points
spilophaeus dark-spotted
Hieracium spilophaeum
spina-christi Christ's thorn
Paliurus spina-christi

CULTIVAR NAMES
Cities of the World

Fuchsia 'City of Adelaide'
Gladiolus 'Amsterdam'
Kniphofia 'Atlanta'
Convallaria majalis 'Berlin Giant'
Clematis integrifolia 'Budapest'
Grevillea 'Canberra Gem'
Hibiscus rosa-sinensis 'Casablanca'
Hemerocallis 'Chicago Apache'
Iris 'Cordoba'
Panicum virgatum 'Dallas Blues'
Persicaria affinis 'Darjeeling Red'
Paeonia lactiflora 'Dresden'
Astilbe 'Düsseldorf'
Clematis 'Duchess of Edinburgh'
Clematis 'Gothenburg'
Iris 'Granada Gold'
Hydrangea macrophylla 'Hamburg'
Pennisetum alopecuroides 'Hameln'
Clematis 'Helsingborg'
Amaryllis belladonna 'Johannesburg'
Canna 'Kansas City'
Fuchsia 'London 2000'
Dianthus 'London Poppet'

Clematis 'Ville de Lyon'
Canna 'Madras'
Lavandula stoechas 'Madrid Blue'
Narcissus 'Mexico City'
Tulipa 'Monte Carlo'
Malus pumila 'Montreal Beauty'
Osteospermum 'Nairobi Purple'
Iris sibirica 'Ottawa'
Clematis 'Etoile de Paris'
Hedera helix 'Pittsburgh'
Canna 'City of Portland'
Primula auricula 'Prague'
Astrantia major 'Roma'
Canna 'Singapore Girl'
Columnea 'Stavanger'
Iris 'Stockholm'
Canna 'Strasbourg'
Canna 'Stuttgart'
Mentha spicata 'Tashkent'
Hydrangea macrophylla 'Tokyo Delight'
Argyranthemum 'Vancouver'
Narcissus 'Verona'
Tulipa 'Yokohama'
Coreopsis verticillata 'Zagreb'

spinosus, spinifer thorny; *spinulifer*
with little spines; *spinalbus* with
white spines
Acanthus spinosus
Androsace spinulifera
Cleome spinosa
Desfontainea spinosa
Eryngium spinalba
Prunus spinosa

spiralis spiral
  Cryptomeria japonica 'Spiralis'
splendens brilliant
  Salvia splendens
splendidum bright, brilliant, splendid
  Helichrysum splendidum
spooneri named for Herman Spooner
  (1878–1976), a botanist at Veitch
  & Sons' nurseries (see page 204)
  Clematis spooneri
sprengeri named for Karl Sprenger
  (1846–1917), a German
  nurseryman working in Italy
  Tulipa sprengeri
spurius false
  Iris spuria
squalidus dirty
  Leptinella squalida
squamatus, squamosus scaly
  Oxalis squamata
squarrosus rough, scurfy; with
  protruding scales
  Chamaecyparis pisifera 'Squarrosa'

STELLATUS

standishii named for John Standish
  (1814–1875), founder of
  Sunningdale Nurseries in Berkshire
  Lonicera standishii
stans upright
  Clematis stans
stellatus stellate, starry, star-shaped
  Magnolia stellata
stellerianus named for Georg Steller
  (1709–1746), a German naturalist
  Artemisia stelleriana
stellulatus rather starlike
  Olearia stellulata
sten-, steno- narrow (see below)
stenantherus with narrow anthers
  Fritillaria stenanthera
stenocephalus with a narrow head
  Ligularia stenocephala

stenophyllus with narrow leaves
Berberis × *stenophylla*
Cistus × *stenophyllus*
Eremurus stenophyllus
Hebe stenophylla
Iris stenophylla

stenopterus with narrow wings
 Pterocarya stenoptera
sternianus, sternii named for Sir
 Frederick Stern (1884–1967),
 British plantsman, gardener
 and author
 Cotoneaster sternianus
 Helleborus × *sternii*
stewartii named for Laurence Stewart
 (1877–1934), Keeper of the Royal
 Botanic Garden Edinburgh
 Digitalis stewartii
stimulosus with stings or prickles
 Phyllostachys stimulosa

stipularis, stipulatus with (large)
 stipules
 Gillenia stipulata
stoechas from the Iles d'Hyères, off
 the French Mediterranean coast
 Lavandula stoechas
stolonifer with rooting runners
 Phlox stolonifera
stragulus, stragulatus mat-forming
 Gentiana stragulata
stramineus like straw, straw-coloured
 Gentiana straminea
strepto- twisted (see below)
streptocarpus with twisted fruits
streptophyllus with twisted leaves

COMMON DESCRIPTIVE TERMS
striatus striped; **striatulus** rather
 stripy
 Aloe striatula
 Geranium sanguineum var. *striatum*
 Polygonatum × *hybridum* 'Striatum'
 Sisyrinchium striatum

strictus upright, erect, tight
 Agave stricta
 Babiana stricta
 Crataegus monogyna 'Stricta'
 Juniperus chinensis 'Stricta'

---

strigillosus
 with short, flattened bristles
strigosus bristly
 Hydrangea aspera subsp. *strigosa*

strobilaceus scaly like a pine-cone
strobilifer bearing cones
struthiopteris ostrich fern
  Matteuccia struthiopteris
studiosorum of scholars
  Androsace studiosorum
stygianus of the River Styx of classical
  mythology
  Euphorbia stygiana
stylosus with a conspicuous or large
  style
  Phuopsis stylosa
styracifluus flowing with gum
  Liquidambar styraciflua
suaveolens fragrant, smelling sweetly
  Mentha suaveolens
sub- almost, somewhat, not
  completely (see below)
subalpinus growing below the
  timber line
  Hebe subalpina
subcrenulatus somewhat scalloped
  Eucalyptus subcrenulata
suberectus somewhat erect
subhirtellus somewhat hairy
  Prunus × subhirtella
subulatus shaped like an awl
  Callistemon subulatus
succulentus juicy, fleshy
  Oxalis succulenta
suecicus Swedish
  Cotoneaster × suecicus
suffruticosus somewhat shrubby
  Buxus sempervirens 'Suffruticosa'

sulphureus sulphur yellow
  Epimedium × versicolor
  'Sulphureum'

COMMON DESCRIPTIVE TERMS
superbus superb
  Chaenomeles × superba
  Leucanthemum × superbum
  Lilium superbum
  Salvia × superba

---

suspensus hanging
  Forsythia suspensa

COMMON DESCRIPTIVE TERMS
sylvaticus, sylvestris of woods,
  or growing wild
  Fagus sylvatica
  Geranium sylvaticum
  Luzula sylvatica
  Malus sylvestris
  Malva sylvestris
  Nicotiana sylvestris
  Nyssa sylvatica
  Pinus sylvestris
  Tulipa sylvestris

---

syriacus from Syria
  Hibiscus syriacus
syringanthus with flowers like *Syringa*
  (lilac)
  Lonicera syringantha
szechuanicus from Sichuan, China
  Betula szechuanica

T

tabularis, tabuliformis flat like a board
   Aeonium tabuliforme
   Blechnum tabulare
taiwanensis, taiwanianus from Taiwan
   Angelica taiwaniana
tamariscifolius with leaves like *Tamarix* (tamarisk)
   Juniperus sabina 'Tamariscifolia'

CULTIVAR NAMES
Famous Writers

DIANTHUS 'JANE AUSTEN'
FUCHSIA 'ALAN AYCKBOURN'
RHODODENDRON 'BALZAC'
NARCISSUS 'BARRETT BROWNING'
ROSA 'ROBERT BURNS'
FUCHSIA 'LORD BYRON'
RHODODENDRON 'CORNEILLE'
HOSTA 'EMILY DICKINSON'
HOSTA 'ROBERT FROST'
CLEMATIS VICTOR HUGO ('EVIPO007')
NARCISSUS 'KEATS'
RHODODENDRON 'SAPPHO'
TULIPA 'SHAKESPEARE'
PELARGONIUM 'SHELLEY'
SEMPERVIVUM 'VIRGIL'
HELIANTHEMUM 'VOLTAIRE'
PELARGONIUM 'WORDSWORTH'

---

tanacetifolius with leaves like *Tanacetum* (tansy)
   Crataegus tanacetifolia
tanguticus from the region of the Tangut people in north-west China
   Daphne tangutica
tardiflorus late-flowering
   Hosta tardiflora
tardivus, tardus late
   Hydrangea paniculata 'Tardiva'
   Tulipa tarda
tarentinus from Taranto in Italy
   Myrtus communis subsp. *tarentina*
tartareus with a crumbly or rough surface
tasmanicus from Tasmania
   Gaultheria tasmanica
tasmanii named for the Dutch explorer Abel Tasman (1603–1659), who discovered Tasmania
   Davallia tasmanii
tataricus, tartaricus from Tatary (Tartary), an old name for a region of central Asia/European Russia
   Lonicera tatarica
tatsienensis from Kangding (Tatsienlu) in Sichuan, China
   Delphinium tatsienense
tauricus from the Crimea
   Asphodeline tauricus
taurinus from the region around Turin (Torino), Italy
taxifolius with leaves like *Taxus* (yew)
   Juniperus taxifolius

Tacamahac *Populus balsamifera*
Tamarisk *Tamarix*
Tansy *Tanacetum vulgare*
Teasel *Dipsacus fullonum*
Tenby daffodil *Narcissus obvallaris*
Texas bluebonnet *Lupinus subcarnosus*
Thorn-apple *Datura stramonium*
Three-cornered leek *Allium triquetrum*
Thrift *Armeria maritima*
Thyme *Thymus*
Tickseed *Coreopsis*
Toad lily *Tricyrtis*
Toadflax *Linaria*
Toothache tree *Zanthoxylum americanum*
Torch lily *Kniphofia*
Traveller's joy *Clematis vitalba*
Tree fern *Dicksonia antarctica*
Tree heath *Erica arborea*
Tree mallow *Lavatera × clementii*
Tree of heaven *Ailanthus altissima*
Trout lily *Erythronium*
Trumpet vine *Campsis*
Tulip tree *Liriodendron tulipifera*
Tupelo *Nyssa*
Turtlehead *Chelone*
Twinflower *Linnaea borealis*

Valerian *Centranthus, Valeriana*
(Vegetables, see page 113)
Venus flytrap *Dionaea muscipula*
Vervain *Verbena officinalis*
Vine *Vitis vinifera*
Violet *Viola*
Viper's bugloss *Echium vulgare*
Virginia creeper *Parthenocissus quinquefolia*

TANSY

tectorum of roofs
Sempervivum tectorum
tenax tenacious, holding fast, tough
Phormium tenax
tenellus rather delicate
Prunus tenella
tenuicaulis with a thin stem
Persicaria tenuicaulis

tenuiculmis with thin culms (stalks)
Carex tenuiculmis

COMMON DESCRIPTIVE TERMS

tenuifolius with thin leaves
Gypsophila tenuifolia
Paeonia tenuifolia
Philadelphus tenuifolius
Pittosporum tenuifolium

WHAT'S IN A NAME?

COLOUR: RED AND ORANGE

calendulinus orange like pot marigold
  (*Calendula*)
capsicinus bright red, like a pepper
  (*Capsicum*)
capucinus orange-red, like nasturtium
  (*Tropaeolum*)
cardinalis cardinal red
carmesinus crimson
carneus flesh-coloured
cerasinus cherry red
coccineus deep carmine red
corallinus coral-coloured
cruentus bloody
erubescens reddening, blushing
erythro- red
flammeus; flammeolus flame-coloured
fuscoruber dark red
igneus fiery red
incarnatus flesh-coloured
kermesinus carmine, purplish red
lateritius brick red
miniatus the colour of red lead
phoeniceus purplish red
puniceus purplish crimson
rhodo- light red
roseus, rosea like a rose, rose-coloured
rubellus light red
ruber red
rubescens reddish, reddening
rubiginosus rusty
rufus, rufescens reddish
sanguineus bloody, dull red
vermiculatus vermilion

tenuis thin, fine, slender
tenuissimus very thin
  Stipa tenuissima
terebinthus turpentine

teres cylindrical
terminalis terminal; of boundaries
  Pachysandra terminalis
ternatus in threes
  Choisya ternata
terniflorus with flowers in threes
ternifolius with three leaves or leaflets
  Gentiana ternifolia
terrestris terrestrial, growing in the
  ground
tessellatus chequered
  Indocalamus tessellatus
testaceus terracotta-coloured
  Carex testacea
testudinarius tortoise-like
tetra- four
tetragonus with four angles
  Chamaecyparis obtusa 'Tetragona
  Aurea'
tetrandrus with four anthers
  Tamarix tetrandra
tetraphyllus with four leaves or
  leaflets
  Oxalis tetraphylla
tetrapterus four-winged
  Sophora tetraptera
teucrioides like *Teucrium*
texanus, texensis from Texas, USA
  Clematis texensis
thapsus former name of a town
  in Sicily
  Verbascum thapsus (great mullein)
thessalus from Thessalia in Greece
  Fritillaria thessala

thibetanus, thibeticus, tibetanus, tibeticus from Tibet
Clematis tibetana
Gentiana tibetica
Helleborus thibetanus
Primula tibetica

thomsonii named for Thomas Thomson (1817–1878), a Scottish doctor and botanist working in India

thunbergii named for the Swedish botanist Carl Thunberg (1743–1828)
Berberis thunbergii
Fritillaria thunbergii
Geranium thunbergii
Hemerocallis thunbergii

thuringiacus from Thuringia in Germany
Lavatera thuringiaca
thymifolius thyme-leaved
Fuchsia thymifolia
thyrsiflorus with flowers in a thyrse (a type of flowerhead, like lilac or horse-chestnut)
Ceanothus thyrsiflorus
thyrsoides like a thyrse (see above)
Campanula thyrsoides
tibetanus, tibeticus (see panel above)

tinctorius used in dyeing
Anthemis tinctoria
Carthamus tinctorius
Genista tinctoria
Impatiens tinctoria
Indigofera tinctoria

tingens dyeing, a dyer
Euonymus tingens
tingitanus from Tangier
Lathyrus tingitanus

tomentosus tomentose, woolly
Cerastium tomentosum
Galactites tomentosa
Paulownia tomentosa
Pelargonium tomentosum
Philadelphus tomentosus
Tilia tomentosa

tommasinianus named for an Italian botanist, Muzio de' Tommasini (1794–1879)
Crocus tommasinianus
topiaria of or for topiary
Hebe topiaria
torminalis curing gripe or colic
Sorbus torminalis
torosus, torulosus cylindrical with bulges or contractions at intervals
torquatus twisted round; with a collar
Helleborus torquatus

tortuosus, tortus, tortilis tortuous, winding
  Fagus sylvatica 'Tortuosa Purpurea'
trachelius curing throat ailments
  Campanula trachelium
tradescantii named for John Tradescant (1570–1638) and his son John (1608–1662), English plant collectors and royal gardeners
  Aster tradescantii
tragophyllus with goat-like leaves
  Lonicera tragophylla
tranquillans calming
  Hibanobambusa tranquillans
transsilvanicus, transylvanicus from Transylvania, Romania
  Hepatica transsilvanica
transitorius passing, transitory
  Malus transitoria
tremulus trembling, quivering
  Populus tremula
triacanthos three-thorned
  Gleditsia triacanthos
triandrus with three stamens
  Narcissus triandrus
triangularis three-angled
  Kniphofia triangularis
tricho- hair (see below)
trichocarpus with hairy fruits
  Populus trichocarpa

trichocaulon with a hairy stem
   Hypericum trichocaulon
trichophyllus with hairy leaves
   Leucojum trichophyllum
trichotomum with divisions in threes
   Clerodendron trichotomum
trichorhizum with hairy rhizomes

COMMON DESCRIPTIVE TERMS
tricolor three-coloured
Convolvulus tricolor
Fagus sylvatica 'Tricolor'
Fuchsia magellanica var. *gracilis*
   'Tricolor'
Hydrangea macrophylla 'Tricolor'
Ipomoea tricolor
Ligustrum lucidum 'Tricolor'
Salvia officinalis 'Tricolor'
Viola tricolor

---

tricornis with three horns
tricuspidatus three-pointed
   Parthenocissus tricuspidata
trifidus divided or cleft into three
   Carex trifida
triflorus three-flowered
   Acer triflorum

COMMON DESCRIPTIVE TERMS
trifolius, trifoliatus three-leaved
Cardamine trifolia
Gillenia trifoliata
Menyanthes trifoliata (bogbean)
Ptelea trifoliata

WHAT'S IN A NAME?
PATTERNS AND MARKINGS

albomaculatus white-spotted
albovariegatus white-variegated
argenteoguttatus silver-spotted
argenteomarginatus silver-edged
argenteovariegatus silver-variegated
astictus without spots
aureomaculatus gold-spotted
aureomarginatus gold-edged
aureoreticulatus gold-veined
erythrostictus with red spots
fasciatus banded, striped
guttatus spotted
irroratus (literally) sprinkled with
  dew; finely spotted
luteovenosus with yellow veins
maculatus, maculifer spotted, blotchy
marginalis, marginatus with a distinct
  margin, edge or border
marmoratus marbled
melanostictus spotted with black
monostictus with one spot
nigropunctatus black-spotted
non-scriptus unmarked
ocellatus, oculatus with an eye
picturatus embellished; variegated
punctatus spotted
reticulatus netted
rubromarginatus edged with red
rufinervis with red veins
spilophaeus dark-spotted
striatus striped;
  striatulus rather stripy
tessellatus chequered
variegatus variegated
viridistriatus green-striped
zebrinus striped
zonalis, zonatus with a distinct band
  of a different colour

trifurcatus three-pronged
 Azorella trifurcata
trilobus, trilobatus three-lobed
 Malus trilobata
trimestris of three months
 Lavatera trimestris

trinervis three-veined
trionum 'of the Wain' (the
 constellation comprising Ursa
 Major and Ursa Minor)
 Hibiscus trionum
tripartitus divided into three parts
 Eryngium × *tripartitum*
tripetalus with three petals
 Magnolia tripetala
triphyllus three-leaved
 Aloysia triphylla
tripinnatus thrice pinnate
triplinervis with three veins
 Anaphalis triplinervis
tripteris three-winged
 Coreopsis tripteris
triquetrus three-cornered
 Allium triquetrum

COMMON DESCRIPTIVE TERMS
tristis sad; **tristulis** rather sad
Betula pendula 'Tristis'
Fritillaria affinis var. *tristulis*
Gladiolus tristis
Pelargonium triste

---

triternatus thrice ternate
 Clematis × *triternata*
trullifolius with scoop-shaped leaves
 Anemone trullifolia
truncatus truncated, cut off
tuberculatus tuberculate, covered
 with small warty nodules
 Echium tuberculatum

tuberosus tuberous
Allium tuberosum (Chinese chives)
Asclepias tuberosa
Geranium tuberosum
Hermodactylus tuberosus
Oxalis tuberosa
Tropaeolum tuberosum

tubiformis tube-shaped
  Fritillaria tubiformis
tubulosus like a small pipe or tube
  Clematis tubulosa
tulipifer tulip-bearing
  Liriodendron tulipifera
tumidissinodus with very swollen
  nodes
  Chimonobambusa tumidissinoda
tuolumnensis from Tuolumne
  County, California (from a Native
  American name)
  Erythronium tuolumnense
turbinatus shaped like a spinning-top
  Aesculus turbinata

turkestanicus from Turkestan
Acer turkestanicum
Dianthus turkestanicus
Salvia turkestanica
Tulipa turkestanica

typhoides like *Typha* (bulrush)
  Kniphofia typhoides

GENUS NAMES
T, U

Taraxacum from Persian *talkh chakok*, bitter herb
Taxodium from Latin *Taxus*, yew, and Greek *eidos*, resemblance
Thunbergia named for Carl Thunberg (1743–1828), Swedish botanist and plant collector
Tiarella diminutive of Greek *tiara*, crown
Tolmiea named for Dr William Tolmie (1812–1886), Scottish doctor and botanist in Canada
Torreya named for Dr John Torrey (1796–1873), leading American botanist
Trachelospermum from Greek *trachelos*, neck, and *sperma*, seed
Trachycarpus from Greek *trachys*, rough, and *karpos*, fruit
Tradescantia named for the John Tradescants, father and son (see page 198)
Tragopogon from Greek *tragos*, goat, and *pogon*, beard
Tricyrtis from Greek *tri-*, three, and *kyrtos*, humped
Trifolium from Latin *tri-*, three, and *folium*, leaf
Trillium from Latin *tri-*, three
Tropaeolum from Latin *tropaeum*, trophy
Tulipa from Turkish *tulbend*, turban
Ursinia named for Johannes Ursinus (1608–1667), German botanical writer
Uvularia from Latin *uvula*, the lobe at the back of the palate

U

ucranicus from the Ukraine
Scabiosa ucranica
uliginosus of marshy, wet places
Salvia uliginosa
ulmarius like *Ulmus* (elm)
Filipendula ulmaria (meadowsweet)

COMMON DESCRIPTIVE TERMS
umbellatus umbelliferous
Butomus umbellatus
Calandrinia umbellata
Elaeagnus umbellata
Iberis umbellata (candytuft)
Ornithogalum umbellatum
Rhaphiolepis umbellata

---

umbraculifer umbrella-bearing
Pinus densiflora 'Umbraculifera'
umbrosus growing in shade
Saxifraga umbrosa
uncinatus hooked or barbed
Clematis uncinata

COMMON DESCRIPTIVE TERMS
undulatus wavy
Asplenium scolopendrium Undulatum Group
Hosta undulata
Mahonia × *wagneri* 'Undulata'
Nerine undulata

unedo 'I eat one'
Arbutus unedo
unguicularis clawed
Iris unguicularis
unifolius with one leaf
Allium unifolium
uniflorus with one flower
Ipheion uniflorum
uplandicus from Uppland, Sweden
Symphytum × *uplandicum*
urbanus, urbium urban, of towns
Saxifraga × *urbium*
urniger with urn-shaped fruits
Eucalyptus urnigera
urophyllus with leaves like an ox's tail
Clematis urophylla
ursinus bearlike; also northern (from the constellation Ursa Major, the Great Bear or the Plough)
Allium ursinum

CULTIVAR NAMES
Strange but True
U, V

PINUS BANKSIANA 'UNCLE FOGY'
JUNCUS 'UNICORN'
PELARGONIUM 'URCHIN'
PHLOX PANICULATA 'UTOPIA'
CLEMATIS 'THE VAGABOND'
PIERIS JAPONICA 'VALLEY VALENTINE'
PINUS FLEXILIS 'VANDERWOLF'S PYRAMID'
ARMERIA MARITIMA 'VINDICTIVE'
PELARGONIUM 'VOODOO'

ARBUTUS UNEDO

urticifolius with leaves like *Urtica*
(nettle)
   Clematis urticifolia
uruguayensis from Uruguay
   Azara uruguayensis
utahensis from Utah, USA
   Penstemon utahensis
usitatissimus very useful
   Linum usitatissimum
utilis useful
   Betula utilis
uva-crispa curly grape
   Ribes uva-crispa (gooseberry)
uvarius like grapes
   Kniphofia uvaria
uva-ursi bear's grape
   Arctostaphylos uva-ursi
uva-vulpis fox's grape
   Fritillaria uva-vulpis

CULTIVAR NAMES
Foreign Expressions
T, U, V

TAUBE (*German*) DOVE
(*HYDRANGEA MACROPHYLLA* 'TAUBE')

TIEFROT (*German*) DEEP RED
(*LOBELIA* 'FAN TIEFROT')

TOITS DE PARIS (*French*)
ROOFS OF PARIS
(*PHLOX PANICULATA* 'TOITS DE PARIS')

TRÈS COUPÉ (*French*) MUCH CUT
(*HEDERA HELIX* 'TRÈS COUPÉ')

TRIOMPHE DE ... (*French*)
TRIUMPH OF ...
(*HELIANTHUS* 'TRIOMPHE DE GAND')

TRISTESSE (*French*) SADNESS
(*SEMPERVIVUM* 'TRISTESSE')

UKIGUMO (*Japanese*)
FLOATING CLOUD
(*ACER PALMATUM* 'UKIGUMO')

VEILCHENBLAU (*German*)
VIOLET BLUE
(*ROSA* 'VEILCHENBLAU')

VEILCHENKÖNIGIN (*German*)
VIOLET QUEEN
(*ASTER AMELLUS* 'VEILCHENKÖNIGIN')

VOIE LACTÉE (*French*)
THE MILKY WAY
(*PHILADELPHUS* 'VOIE LACTÉE')

VORLÄUFER (*German*)
FORERUNNER
(*MISCANTHUS SINENSIS* 'VORLÄUFER')

VUURBAAK (*Dutch*) LIGHTHOUSE
(*HYACINTHUS ORIENTALIS* 'VUURBAAK')

V

vacciniifolius with leaves like
Vaccinium
  Persicaria vacciniifolia
vagans, vagensis wandering; widely
distributed
  Erica vagans
  Sorbus × *vagensis*
valdivianus from Valdivia in Chile
  Berberis valdiviana
valentinus from Valencia, Spain
  Coronilla valentina
validus strong, robust, powerful
  Lobelia valida
vanhouttei named for a Belgian
nurseryman, Louis Van Houtte
(1810–1876)
  Spiraea × *vanhouttei*
variabilis, varians, variatus variable
  Erodium × *variabile*

variegatus variegated
Aloe variegata
Iris pallida 'Variegata'
Miscanthus sinensis 'Variegatus'
Pleioblastus variegatus
Rhamnus alaternus 'Variegata'

variifolius with variable leaves
  Eryngium variifolium

veitchii, veitchianus named for
Veitch & Sons, a famous 19th-
century nursery founded by John
Veitch (1752–1839) and run by
five generations of the Veitch
family in Exeter and London
Abies veitchii
Ligularia veitchii
Paeonia veitchii
Sasa veitchii
Saxifraga veitchiana

velatus covered, concealed
vellerius fleecy
velutinus velvety
  Fraxinus velutina
venenatus, venenosus poisonous
venetus from Venice
  Lathyrus venetus
venosus veined, prominently veined
  Clematis 'Venosa Violacea'
ventricosus (literally) pot-bellied;
with a swelling
  Hosta ventricosa

venustus lovely, charming, pleasing
Adiantum venustum
Eryngium venustum
Filipendula rubra 'Venusta'
Fuchsia venusta
Hosta venusta
Penstemon venustus

WHAT'S IN A NAME?

PLACES

MANY LATIN AND LATINIZED PLACE-NAMES, SUCH AS *AMERICANUS*, *ITALICUS* AND *GERMANICUS*, NEED NO EXPLANATION. THE FOLLOWING LIST GIVES A SAMPLE OF DESCRIPTIVE TERMS DERIVED FROM SOME LESS OBVIOUS ONES.

aegypticus, aegyptiacus Egypt

aethiopicus, aethiopis Ethiopia; also used to mean South African

afer, afra, afrum; africanus Africa

anatolicus Anatolia, Turkey

ancyrensis Ankara, Turkey

asturiensis Asturias, Spain

atticus Attica or Athens, Greece

baeticus Andalucia, Spain

berolinensis Berlin

bonariensis Buenos Aires, Argentina

bucarius, bucharicus Bokhara, Turkestan

byzantinus Istanbul (formerly Byzantium), Turkey

calabricus Calabria, Italy

canadensis Canada or the north-eastern USA

canariensis Gran Canaria or the Canary Islands

cantabricus Cantabria in Spain

capensis the Cape (of Good Hope), meaning South Africa

caucasicus Caucasia, former Soviet Union

cornubiensis Cornwall, England

creticus Crete

cyprius Cyprus

damascenus Damascus, Syria

etruscus Tuscany, Italy

fennicus Finland

florentinus Florence, Italy

fluminensis Rio de Janeiro

formosanus Taiwan (Formosa)

georgicus Georgia (Caucasus)

groenlandicus Greenland

ibericus the Iberian peninsula

illyricus Illyria (the Adriatic coast of Croatia and Dalmatia)

japonicus, niponicus, nipponicus Japan

libani, libanotis, libanoticus Lebanon

lusitanicus Portugal

lutetianus Paris

magellanicus the southern tip of South America, around the Straits of Magellan

monacensis Munich, Germany

monspeliensis, monspessulanus Montpellier, France

narbonensis Narbonne, France

norvegicus Norway

novae-zelandiae New Zealand

numidicus Algeria

oxonianus, oxoniensis Oxford, England

pedemontanus Piedmont, Italy

pekinensis Beijing, China

ponticus Pontus, Asia Minor

ruthenicus Ruthenia, a region of Russia (and used more generally to mean Russian)

sabatius Savona, on the Ligurian coast of north-west Italy

sarniensis Guernsey, Channel Islands

siculus Sicily

sinensis, sinicus China

sitchensis Sitka, Alaska

syriacus Syria

valentinus Valencia, Spain

tarentinus Taranto, Italy

tauricus the Crimea

yedoensis, yesoensis, yezoensis Tokyo

veris, vernalis, vernus of spring

Crocus vernus

Gentiana verna

Hamamelis vernalis

Lathyrus vernus

Leucojum vernum

Omphalodes verna

Primula veris

Pulsatilla vernalis

vernicosus varnished
   Eucalyptus vernicosa
verrucosus warty
   Euonymus verrucosus
verruculosus with little warts
   Berberis verruculosa

versicolor in various colours

Brugmansia versicolor

Epimedium × *versicolor*

Geranium versicolor

Iris versicolor

Lupinus versicolor

Oenothera versicolor

Oxalis versicolor

verticillatus verticillate; in whorls
   Coreopsis verticillata

WHAT'S IN A NAME?

COLOUR: GREEN

aeruginosus blue-green
   (the colour of verdigris)
atrovirens dark green
chlorinus yellow-green
chloro- clear green
elaio- olive green
flavovirens yellowish green
glaucus, glaucescens dull green
   to greyish blue
olivaceus olive green, brownish
   green
prasinus leek green
smaragdinus emerald green, clear
   bright green
thalassicus sea green
virens verdant green
virescens greenish, becoming green
viridescens greenish
viridis green (general)
viridissimus very green
viridulus greenish
xanthochlorus yellowish green

PRIMULA VERIS

verus true; standard
　Aloe vera
vescus small, feeble
　Fragaria vesca
vesicarius relating to the bladder;
　a remedy for bladder ailments
　Eruca vesicaria
vestitus covered or clothed
　Polystichum vestitum
vialii named for Paul Vial (1855–
　1917), a French missionary
　Primula vialii
viburnifolius with leaves like
　Viburnum
viburnoides like *Viburnum*
　Pileostegia viburnoides
victoriae-reginae Queen Victoria
　Agave victoriae-reginae
vigilis watchful, alert
　Diascia vigilis

COMMON DESCRIPTIVE TERMS
villosus softly hairy
Androsace villosa
Hydrangea aspera Villosa Group
Mentha × *villosa*
Pennisetum villosum
Soldanella villosa

---

vilmorinii named for Vilmorin-
　Andrieux, a long-established
　French nursery and seed
　supplier
　Sorbus vilmorinii

CULTIVAR NAMES:
Personal Names
T, U, V

FUCHSIA 'TAFFY'
LYCHNIS 'TERRY'S PINK'
NARCISSUS 'THALIA'
VITIS VINIFERA 'THERESA'
DIANTHUS 'THOMAS'
KNIPHOFIA 'TIMOTHY'
GALANTHUS NIVALIS 'TINY TIM'
NARCISSUS 'TOBY'
TULIPA 'UNCLE TOM'
PRIMULA 'TONY'
NARCISSUS 'TRACEY'
GERANIUM × OXONIANUM
'TREVOR'S WHITE'
HEBE 'TRIXIE'
PELARGONIUM 'TRUDIE'
FUCHSIA 'TRUDY'
MISCANTHUS SINENSIS 'UNDINE'
HEDERA HELIX 'URSULA'
DIASCIA 'LADY VALERIE'
CLEMATIS 'VANESSA'
CLEMATIS 'VERA'
CLEMATIS 'VERONICA'S CHOICE'
SALVIA FARINACEA 'VICTORIA'
CLEMATIS 'VIOLA'
LIMONIUM PLATYPHYLLUM
'VIOLETTA'

viminalis, vimineus with long, thin
　shoots; like osiers
　Salix viminalis (osier)
vinciflorus with flowers like *Vinca*
vinealis of vines or vineyards
　Allium vineale

VITIS VINIFERA

vinifer wine-producing
  Vitis vinifera (grape vine)

violaceus violet
Clematis 'Venosa Violacea'
Fabiana imbricata f. *violacea*
Festuca violacea
Passiflora × *violacea*

virens green
virescens greenish, becoming green
  Galanthus nivalis 'Virescens'
virgatus twiggy
  Corokia × *virgata*

virginianus, virginicus from
  Virginia, USA
Chionanthus virginicus
Hamamelis virginiana
Itea virginica
Juniperus virginiana
Magnolia virginiana
Mertensia virginica
Persicaria virginiana
Physostegia virginiana
Veronicastrum virginicum

viridapice with green tips
  Galanthus nivalis 'Viridapice'

viridiflorus with green flowers
Aquilegia viridiflora
Callistemon viridiflorus
Digitalis viridiflora
Galtonia viridiflora

viridis green; **viridissimus** very
  green; **viridescens** becoming green;
  viridulus rather green
Carex viridula subsp. *viridula*
Crataegus viridis 'Winter King'
Disporum viridescens
Forsythia viridissima
Helleborus viridis
Lavandula viridis
Salvia viridis var. *comata* (*Salvia
  horminum*)

viridistriatus green-striped
  Pleioblastus viridistriatus
viscosus sticky, viscous
  Rhododendron viscosum
vitalba white vine
  Clematis vitalba
vitellinus egg-yolk yellow
  Salix alba subsp. *vitellina*
viticella little vine
  Clematis viticella
vitifolius with leaves like *Vitis* (vine)
  Abutilon × *vitifolium*
volubilis winding, revolving
vomitorius causing vomiting
  Ilex vomitoria
vulcanicus of volcanoes, growing
  on a volcano
  Fuchsia vulcanica

COMMON DESCRIPTIVE TERMS

vulgaris common
Aquilegia vulgaris (columbine)
Calluna vulgaris (heather)
Echium vulgare (viper's bugloss)
Filipendula vulgaris (dropwort)
Foeniculum vulgare (fennel)
Leucanthemum vulgare (oxeye daisy)
Origanum vulgare (wild marjoram)
Polypodium vulgare (common
  polypody)
Primula vulgaris (primrose)
Pulsatilla vulgaris (Pasque flower)
Syringa vulgaris (lilac)
Thymus vulgaris (common thyme)

GENUS NAMES
V,W,X,Y,Z

Valeriana from Latin *valere*,
  to be well
Veronicastrum from *Veronica* and
  Latin *-astrum*, which denotes an
  incomplete resemblance
Vesicaria from Latin *vesica*, bladder
Vinca from Latin *vincire*, to wind
  around or bind
Waldsteinia named for Count
  Franz Waldstein-Wartenburg
  (1759–1823), Austrian botanist
  and author
Washingtonia named for George
  Washington (1732–1799), first
  President of the USA
Weigela named for Christian
  Weigel (1748–1831), German
  botanist
Wisteria named for Caspar Wistar
  (1761–1818), American professor
Xerochrysum from Greek *xeros*, dry,
  and *chrysos*, gold
Yucca from a Caribbean word
  meaning cassava or manioc
Zantedeschia named for Francesco
  Zantedeschi (1797–1873), Italian
  priest and physicist
Zephyranthes from Greek *zephyros*,
  the west wind, and *anthos*, flower
Zinnia named for Johann Zinn
  (1727–1759), German botanist

vulnerarius of wounds; for wounds
  Anthyllis vulneraria
vulpinus of foxes
  Carex vulpina

W

waldsteinii named for an Austrian botanist, Count Franz Waldstein-Wartenburg (1759–1823)
Cardamine waldsteinii

COMMON DESCRIPTIVE TERMS
wallichianus, wallichii named for Nathaniel Wallich (1786–1854), a Danish botanist and doctor who specialized in the flora of India
Dryopteris wallichiana
Euphorbia wallichii
Geranium wallichianum
Lilium wallichianum
Meconopsis wallichii
Persicaria wallichii
Pinus wallichiana
Selinum wallichianum
Strobilanthes wallichii

---

wardii (see panel above right)
warleyensis named for Warley Place, Essex, once famous as the large and lavish garden of Ellen Willmott (see *willmottianus*, page 212)
Epimedium × warleyense
warscewiczii named for a Polish plant collector, Joseph Warsczewicz (1812–1866)
Canna warscewiczii

FRANK KINGDON WARD (1885–1958)

Carrying the tradition of the great plant hunters into the 1950s, Kingdon Ward was the last in a 300-year line. Beginning when he was a schoolmaster in Shanghai, he explored the remoter reaches of the Far East for nearly 50 years, introducing many rhododendrons, lilies, gentians, primulas and the beautiful Tibetan blue poppy, *Meconopsis betonicifolia*. Plants commemorating him include *Cotoneaster wardii* and *Rhododendron wardii*. *Primula florindae* and *Lilium mackliniae* are named for his first and second wives respectively.

watereri named for members of several generations of the Waterer family – owners, from the 18th to the 20th century, of a leading nursery firm based at Bagshot and Woking in Surrey
Cotoneaster × watereri

Wake-robin *Trillium*
Wallflower *Erysimum*
Walnut *Juglans*
Water avens *Geum rivale*
Water lily
  Nymphaea
Water mint
  Mentha aquatica
Water soldier
  Stratiotes aloides
Water violet
  Hottonia palustris
Wattle *Acacia*
Wayfaring-tree
  Viburnum lantana
Wellingtonia
  Sequoiadendron
  giganteum
Welsh poppy
  Meconopsis
  cambrica
Whitebeam
  Sorbus aria
Wild marjoram
  Origanum vulgare
Wild service tree *Sorbus torminalis*
Willow *Salix*
Windflower *Anemone*
Wingnut *Pterocarya*
Winter aconite *Eranthis hyemalis*
Winter heliotrope *Petasites fragrans*

WALNUT

Winter jasmine *Jasminum nudiflorum*
Winter sweet *Chimonanthus praecox*
Winter's bark *Drimys winteri*
Winter-flowering cherry *Prunus*
  × *subhirtella*
  'Autumnalis'
Wire-netting bush
  Corokia cotoneaster
Witch-hazel
  Hamamelis
Wood anemone
  Anemone nemorosa
Wood sorrel *Oxalis*
  acetosella
Woodbine *Lonicera*
  periclymenum
Woodruff *Asperula*;
  Galium odoratum
Woodrush *Luzula*
  sylvatica
Wormwood
  Artemisia absinthium
Yarrow *Achillea*
  millefolium
Yellow archangel *Lamium galeobdolon*
Yellow loosestrife *Lysimachia vulgaris*
Yellow rattle *Rhinanthus minor*
Yellow wood *Cladrastis lutea*
Yew *Taxus baccata*
Yew, Irish *Taxus baccata* 'Fastigiata'
Yulan *Magnolia denudata*

weyrichii named for Heinrich
  Weyrich (1828–1868), a German
  botanist
wherryi named for American scientist
  Edgar Wherry (1885–1982)
  Tiarella wherryi

whittallii named for Edward
  Whittall (1851–1917), British
  Consul and merchant in Turkey,
  and a keen amateur botanist and
  plant collector
  Fritillaria whittallii

williamsii named for one of several botanists and collectors called Williams; perhaps the British botanist John Williams (1915–1991)
  Camellia × *williamsii*
willmottianus named for Miss Ellen Willmott (1858–1934), a keen amateur plantswoman who created an extravagant garden at Warley Place, Essex
  Ceratostigma willmottianum
wilsonii (see panel below)
winteri named for Captain Winter, an Elizabethan seafarer
  Drimys winteri
wintonensis from Winchester, Hampshire (usually applied to plants raised by Hillier Nurseries)
  × *Halimiocistus wintonensis*

CULTIVAR NAMES
Personal Names
W, X, Y, Z
PRIMULA 'WANDA'
GALANTHUS PLICATUS 'WENDY'S GOLD'
CAMELLIA JAPONICA 'WILAMINA'
ROSA 'WILHELM'
CARDAMINE PRATENSIS 'WILLIAM'
CLEMATIS 'WILLY'
STREPTOCARPUS 'WINIFRED'
LINARIA 'WINIFRID'S DELIGHT'
MALUS DOMESTICA (APPLE) 'WINSTON'
CHAMAECYPARIS LAWSONIANA 'YVONNE'
FUCHSIA 'ZARA'
PRIMULA AURICULA 'LADY ZOË'
SEMPERVIVUM 'ZORBA'

ERNEST HENRY WILSON (1876–1930)
Ernest Henry Wilson made his first trip to China in 1899 for the nursery of Veitch & Sons (see page 204). In a life full of adventure, Wilson later travelled widely in China and became one of the most prolific of all plant hunters, introducing hundreds of species to the Western world. In 1927 he became Keeper of the Arnold Arboretum in Massachusetts, but died three years later in a car accident. 'Chinese' Wilson introduced such species as *Lilium regale* and *Kolkwitzia amabilis* (his favourite). Plants commemorating him include *Ilex* × *altaclerensis* 'Wilsonii', *Iris wilsonii* and *Primula wilsonii*.

WALDFEE (*German*) WOOD FAIRY
(*FUCHSIA* 'WALDFEE')

WÄLDLER (*German*)
WOODLANDER
(*LUZULA SYLVATICA* 'WÄLDLER')

WARSZAWSKA NIKE (*Polish*)
VICTORY OF WARSAW
(*CLEMATIS* 'WARSZAWSKA NIKE')

WEISSE GLORIA (*German*)
WHITE GLORIA
(*ASTILBE* 'WEISSE GLORIA')

WEISSE SCHWAN (*German*)
WHITE SWAN
(*PULSATILLA VULGARIS* 'WEISSE SCHWAN')

WEISSER ZWERG (*German*)
WHITE DWARF
(*IBERIS SEMPERVIRENS* 'WEISSER ZWERG')

WESTFALEN (*German*)
WESTPHALIA
(*HYDRANGEA PANICULATA* 'WESTFALEN')

WETTERFAHNE (*German*)
WEATHER VANE
(*MISCANTHUS SINENSIS* 'WETTERFAHNE')

WINDSPIEL (*German*) WIND PLAY
(*MOLINIA CAERULEA* SUBSP. *CAERULEA* 'WINDSPIEL')

WINTERMÄRCHEN (*German*)
WINTER FAIRY TALES
(*BERGENIA* 'WINTERMÄRCHEN')

WUNDER VON STÄFA (*German*)
WONDER OF STÄFA ·
(*ASTER* × *FRIKARTII* 'WUNDER VON STÄFA')

YUKIKOMACHI (*Japanese*)
MORNING CLOUD
(*CLEMATIS* 'YUKIKOMACHI')

ZIGEUNERKNABE (*German*)
GYPSY BOY
(*ROSA* 'ZIGEUNERKNABE')

ZIMBELSTERN (*German*)
CYMBAL STAR
(*HELENIUM* 'ZIMBELSTERN')

ZWARTE SNOR (*Dutch*)
BLACK MOUSTACHE
(*FUCHSIA* 'ZWARTE SNOR')

ZWARTKOP (*Dutch*) BLACK HEAD
(*AEONIUM* 'ZWARTKOP')

ZWEIWELTENKIND (*German*)
CHILD OF TWO WORLDS
(*ARUNCUS DIOICUS* 'ZWEIWELTENKIND')

ZWERGELEFANT (*German*)
BABY ELEPHANT
(*MISCANTHUS SINENSIS* 'ZWERGELEFANT')

wisleyensis named for the Royal Horticultural Society's garden at Wisley, Surrey
Gaultheria × *wisleyensis*

wulfenii, wulfenianus named for an Austrian botanist, Franz von Wulfen (1728–1805)
Euphorbia characias subsp. *wulfenii*

X, Y, Z

xanth- yellow (see below)

xanthinus yellow
  Rosa xanthina

xanthocarpus with yellow fruit
  Sorbus aucuparia var. *xanthocarpa*

xanthocodon yellow bell
  Rhododendron cinnabarinum subsp. *xanthocodon*

xanthochlorus yellowish green
  Alchemilla xanthochlora

xiphium gladiolus; sword
  Iris xiphium

xylo- wood (see below)

xylocanthus with woody spines

xylocarpus with woody fruit
  Sinojackia xylocarpa

yakushimanus from Yaku-shima, a small island in southern Japan
  Rhododendron yakushimanum

COMMON DESCRIPTIVE TERMS

yedoensis, yesoensis, yezoensis from Tokyo
Cardamine yezoensis
Chrysanthemum yezoense
Geranium yesoense
Polemonium yezoense
Prunus × *yedoensis*

---

yuccifolium with leaves like *Yucca*
  Eryngium yuccifolium

COMMON DESCRIPTIVE TERMS

yunnanensis from Yunnan, China
Cardiocrinum giganteum var. *yunnanense*
Hedychium yunnanense
Lysimachia yunnanensis
Osmanthus yunnanensis
Pinus yunnanensis
Rhododendron yunnanense
Syringa yunnanensis

CULTIVAR NAMES
Strange but True
W, X, Y, Z

NARCISSUS 'WALDORF ASTORIA'
PELARGONIUM 'WAYWARD ANGEL'
RHODODENDRON 'WEE BEE'
DIERAMA 'WESTMINSTER CHIMES'
GAURA LINDHEIMERI 'WHIRLING BUTTERFLIES'
FUCHSIA 'WIDOW TWANKY'
NARCISSUS 'WITCH DOCTOR'
CORNUS KOUSA 'WOLF EYES'
RHODODENDRON 'WOMBAT'
ROSA X-RATED ('TINX')
NARCISSUS 'XIT'
HEMEROCALLIS 'YABBA DABBA DOO'
RANUNCULUS FICARIA 'YAFFLE'
IRIS 'YO-YO'
LINARIA × *DOMINII* 'YUPPIE SURPRISE'
IRIS 'ZINC PINK'
IRIS 'ZIPPER'

WHAT'S IN A NAME?

NATURAL FEATURES

aetnensis Mount Etna (Sicily)

aleuticus Aleutian Islands

alleghaniensis Allegheny Mountains (eastern USA)

altaicus Altai Mountains (Siberia/ Kazakhstan/Mongolia/China)

amanus Amanus range (Turkey)

amazonicus River Amazon

amurensis Heilong Jiang (Amur river) (Russia/China)

andinus the Andes

antarcticus Antarctica

apenninus the Apennines (Italy)

atlanticus Atlas Mountains (N. Africa)

azoricus the Azores

balticus the area around the Baltic Sea

carpaticus Carpathian Mountains

cevennensis the Cévennes (France)

chathamicus the Chatham Islands

chiloensis Chiloé, an island off Chile

coum Kos (a Greek island)

dolomiticus the Dolomites (Italy)

emodensis, emodi the Himalayas

faeroensis Faeroe Isles

fuegianus Tierra del Fuego

karataviensis, karatavicus Karatau Mountains (Kazakhstan)

madagascariensis Madagascar

maderensis Madeira

megapotamicus (literally) big river: the Rio Grande (Brazil)

nootkatensis Nootka Sound (Canada)

olbius, olbiensis Iles d'Hyères (off the Mediterranean coast of France)

omeianus, omeiensis Emei Shan (Mount Omei, in Sichuan, China)

pyrenaicus, pyrenaeus the Pyrenees

rhodopensis Rhodope Mountains (Bulgaria)

sachalinensis Sakhalin island (east Asia)

saluenensis Nu Jiang (Salween river) (Burma/China)

scilloniensis Isles of Scilly

tasmanicus Tasmania (Australia)

---

zabelii named after Hermann Zabel (1832–1912), a German tree expert
  Eryngium × *zabelii*

zawadskii named after Alexander Zawadski (1798–1868), an Austrian scientist

---

COMMON DESCRIPTIVE TERMS

zebrinus striped
  Malva sylvestris 'Zebrina'
  Miscanthus sinensis 'Zebrinus'
  Musa acuminata 'Zebrina'
  Pinus wallichiana 'Zebrina'

zetlandicus from the Shetland Islands

zeylanicus from Sri Lanka (Ceylon)
  Rhododendron arboreum subsp. *zeylanicum*

zonalis, zonatus with a distinct band of a different colour
  Pelargonium zonale

zoysii named after Karl von Zoys (1756–1800), an Austrian botanist
  Campanula zoysii

zuluensis from the former kingdom of Zululand in southern Africa
  Plectranthus zuluensis

GLOSSARY

A

Acuminate With a long point, tapering

Adpressed Lying close and flat against a surface

Alternate (of leaves) Borne singly at each node on opposite sides of the stem

Anther The pollen-bearing part of the stamen

Articulate Jointed

Auricle An ear-shaped projection or appendage

Axil The angle formed by a leaf or lateral branch with the stem, or of a vein with the midrib

B

Bifurcate Divided into two branches

Bipinnate Twice pinnate

Blade The expanded part of a leaf or petal

Bract A modified, usually reduced leaf at the base of a flower stalk

Bulbil A small bulb-like structure

C

Calyx The outer part of the flower, the sepals

Carpel Individual female part of a flower, comprising style, stigma and ovary

Catkin A normally dense spike or spike-like raceme of tiny, scaly-bracted flowers or fruits

Compound Composed of two or more similar parts

Corolla The inner, normally conspicuous part of a flower, the petals

Corymb A flat-topped or dome-shaped flowerhead with the outer flowers opening first

Crenate Toothed with shallow, rounded teeth; scalloped

Cultivar Distinct form not considered to warrant full botanical recognition, selected from either cultivated or wild plants and maintained in cultivation by propagation (example: *Heuchera* 'Plum Pudding')

Cuneate Wedge-shaped

Cuspidate Abruptly sharp pointed

Cyme A flat-topped or dome-shaped flowerhead with the inner flowers opening first

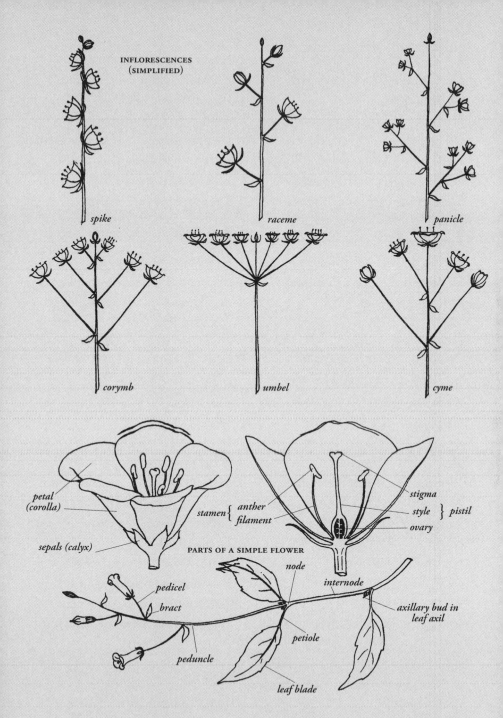

INFLORESCENCES
(SIMPLIFIED)

spike

raceme

panicle

corymb

umbel

cyme

petal (corolla)

sepals (calyx)

stamen { *anther* *filament*

stigma

style } *pistil*

ovary

PARTS OF A SIMPLE FLOWER

pedicel

bract

peduncle

node

internode

axillary bud in leaf axil

petiole

leaf blade

D

Decumbent Reclining, with ascending tips

Decurrent Extending down the stem

Decussate With x-shaped leaves at right angles to those below

Dehiscent Opening as seed vessels

Dioecious Bearing male and female flowers on different plants

Dissected Divided into many narrow segments

Distal Situated away from the point of attachment

Double (of flowers) With more than the usual number of petals, often with the style and stamens changed to petals

E

Elliptic Widest in the middle, narrowing equally at both ends

Entire Undivided and without teeth

Exstipulate Without stipules

F

Fastigiate With branches erect and close together

Fibrillose Having (or composed of) thin fibres, like the finest roots of a plant

Filament The stalk of a stamen

Fimbriate Fringed

Fistular Hollow like a pipe

Florets Small, individual flowers of a dense inflorescence

Forma (form; f.) Slightly differing variant of a species, ranked below varietas (var.). See also page 13

Fusiform Spindle-shaped

G

Genus In plant classification, the primary category, between family and species, grouping together species that have many characteristics in common. See also pages 11–12

Glabrous Hairless

Glaucous Covered with a bloom, bluish-white or silvery

H, I

Hybrid A plant resulting from a cross between different species

Inflorescence The flowering part of the plant

Internode The portion of stem between two nodes or joints

Involucre A whorl of bracts surrounding a flower or flower cluster

L

Lanceolate Lance-shaped

Lateral On or at the side

Leaflet Part of a compound leaf

Linear Long and narrow with nearly parallel margins

Lip One of the parts of an unequally divided flower

Lobe Any protruding part of an organ (as in leaf, corolla or calyx)

M, N, O

Midrib The central vein or rib of a leaf

Node The place on the stem where the leaves are attached; the joint

Oblong Longer than broad, with nearly parallel sides

Obovate Inversely ovate

Opposite (leaves) Borne two to each node, opposite each other.

Oval Broadest in the middle

Ovary The basal 'box' part of the pistil, containing the ovules

Ovate Broadest below the middle (like a hen's egg)

P

Palmate Lobed or divided in hand-like fashion, usually 5- or 7-lobed

Panicle A branching raceme

Pedicel The stalk of an individual flower in an inflorescence

Peduncle The stalk of a flower cluster or of a solitary flower

Perfoliate Having pairs of opposite leaves fused at the base, so the stem appears to pass through them

Perianth The calyx and corolla together; also used for a flower in which there is no distinction between corolla and calyx

Perule The covering of a seed or leaf-bud

Petal One of the separate segments of a corolla

Petiole The leaf-stalk

Pinnate With leaflets arranged on either side of a central stalk

Pistil The female reproductive part of a flower, comprising one or more carpels (ovary, style and stigma)

Pollen Spores or grains contained in the anther, containing the male element

R

Raceme A simple elongated inflorescence with stalked flowers

Rib A prominent vein in a leaf

Runner A trailing shoot taking root at the nodes

S

Scale A minute leaf or bract, or a flat gland-like appendage on the surface of a leaf, flower or shoot

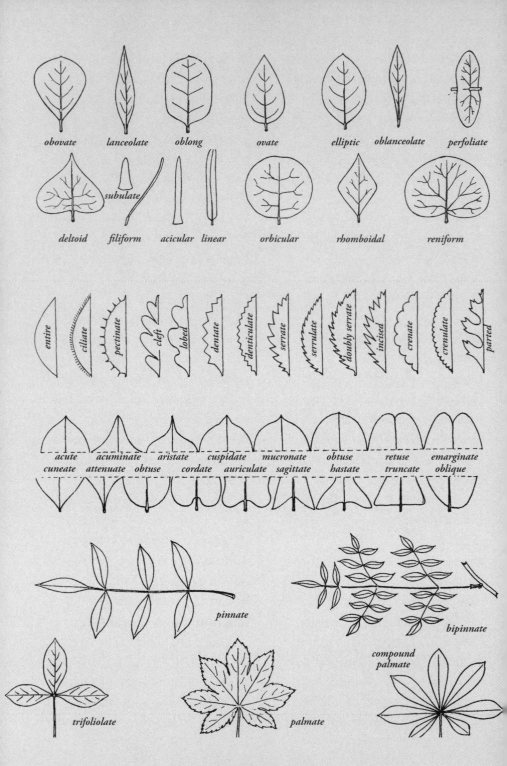

obovate   lanceolate   oblong   ovate   elliptic   oblanceolate   perfoliate

deltoid   subulate   filiform   acicular   linear   orbicular   rhomboidal   reniform

entire   ciliate   pectinate   cleft   lobed   dentate   denticulate   serrate   serrulate   doubly serrate   incised   crenate   crenulate   parted

acute   acuminate   aristate   cuspidate   mucronate   obtuse   retuse   emarginate
cuneate   attenuate   obtuse   cordate   auriculate   sagittate   hastate   truncate   oblique

pinnate

bipinnate

compound
palmate

trifoliolate   palmate

Scape A leafless stem of a flower or inflorescence

Secund Arranged on one side only

Sepal One of the segments of a calyx

Sessile Not stalked

Simple (of a leaf) Not compound; (of an inflorescence) unbranched

Sorus (plural sori) A cluster of spore-producing organs on the underside of a fern frond

Spathe A hoodlike bract surrounding a flower spike

Species In plant classification, a subdivision of a genus, grouping together individual plants with certain distinguishing characteristics. See also pages 11–12

Spine A sharp, pointed end of a branch or leaf

Stamen The male organ of a flower comprising filament and anther

Stigma The summit of the pistil, which receives the pollen, often sticky or feathery

Stipule Appendage (normally two) at the base of some petioles

Stolon A shoot at or below the surface of the ground that produces a new plant at its tip

Style The middle part of the pistil, often elongated, between the ovary and stigma

Subspecies (subsp.) In plant classification, a rank below species but above varietas (variety) and forma (form). See also page 13

T

Tendril A twining thread-like appendage

Terminal Located at the end of a stem or shoot

Ternate In threes

Thyrse A type of contracted flower panicle, like lilac or horse-chestnut

Tuberculate Having small wartlike projections

U, V, W

Umbelliferous Bearing umbels, flat-topped inflorescences in which the pedicels or peduncles all arise from the same point

Varietas (variety; var.) A variant of a species, occurring in the wild, and ranked below subspecies and above forma. See also page 13

Venation The arrangement of the veins

Verticillate Arranged in a whorl or ring

Whorl Three or more flowers or leaves arranged in a ring

INDEX

ACKNOWLEDGEMENTS

David & Charles and OutHouse Publishing would like to thank the following organizations and individuals for their assistance in the preparation of this book: the Royal Horticultural Society for their help in finding and providing the portraits of plant hunters and plantsmen; the staff of the RHS Lindley Library for their patient and friendly help with research; Hillier for permission to use illustrations and botanical information from the *Hillier Gardener's Guide to Trees & Shrubs*, and for the photograph of Sir Harold Hillier; Ian Smith and Mike Park for assistance in sourcing illustrations; Dover Press for permission to use illustrations from the *Handbook of Plant and Floral Ornament from Early Herbals*; Yukari Paul for help in translating Japanese cultivar names; and Christopher Gordon for valuable assistance with proofreading.